Data Science, Machine Intelligence, and Law

Volume 1

Data-driven technologies, such as AI systems and robotics, have the potential to improve our lives, our society and our economy. At the same time, the development and use of these technologies raise unresolved legal and ethical issues. Our new peer reviewed book series brings together works from different legal fields, dealing with the legal, social and economic impact of these technologies, exploring on the one hand the opportunities for law and legal practice, and on the other hand, the ethical and legal frameworks they require. The series especially welcomes book proposals dealing with the legal and/or ethical aspects of data science and the emerging discipline of machine intelligence, which encompasses perception (sensing), AI (planning) and robotics (acting). In this way, our book series aims to contribute to an informed dialogue across the various disciplines about the chances and risks of data-driven systems.

More information about this series at http://www.springer.com/series/16609

Martin Ebers • Marta Cantero Gamito
Editors

Algorithmic Governance and Governance of Algorithms

Legal and Ethical Challenges

Springer

Editors
Martin Ebers
Humboldt University
of Berlin, Law Faculty
Berlin, Germany

University of Tartu,
School of Law
Tartu, Estonia

Marta Cantero Gamito
CUNEF, Law Department
Madrid, Spain

University of Tartu,
School of Law
Tartu, Estonia

ISSN 2730-5899 ISSN 2730-5902 (electronic)
Data Science, Machine Intelligence, and Law
ISBN 978-3-030-50558-5 ISBN 978-3-030-50559-2 (eBook)
https://doi.org/10.1007/978-3-030-50559-2

This Springer imprint is published by the registered company Springer Nature Switzerland AG.
The registered company address is: Gewerbestrasse 11, 6330 Cham, Switzerland

Foreword

Algorithmic systems are increasingly being used as part of decision-making processes in both public and private sectors, with potentially significant consequences for individuals, organisations and societies as a whole. It is no surprise that many types of decisions can be made faster and more efficiently using computational algorithms. However, at the same time, the scale and complexity of the decisions that are being delegated to machines together with the autonomic nature of these decisions are raising concerns about the transparency, fairness, privacy and lack of explanation of these applications. Hardly a day goes by without news of yet another case of manipulation, biases, censorship, social discrimination, violations of privacy and property rights related to the application of algorithmic systems.

The lack of transparency which prevents meaningful scrutiny and accountability is a significant concern when these systems are applied as part of decision-making processes that can have a considerable impact on people's self-determination and human rights. Moreover, the adaptive nature of many artificial intelligence (AI) algorithms together with their opacity also affects privacy and non-discrimination: if we cannot understand how the machine has reached a decision and there are no guarantees that equal conditions lead to equal results, there can be no guarantee of fairness and equal treatment.

Organisations and governments across the world are defining governance strategies to provide appropriate safeguards to protect individuals and society. As an example, High Level Expert Group on AI appointed by the European Commission has defined concrete recommendations for trustworthy AI that aim exactly at this. However, scholar, multidisciplinary research that can provide a sound basis for understanding the impact of algorithms and the possibilities for regulation is often still lacking.

As such, this present volume is an extremely welcome contribution to the advance of state of the art on algorithmic governance. The papers in this volume address these challenges and therefore contribute to shape the discussion about the current and future national and international legal framework for AI and robotics. Starting with a discussion of fundamental principles, this volume further presents work on responses to behavioural profiling, means to deal with algorithmic opacity, labour and societal impact and challenges and different national approaches to privacy and security. It also includes a discussion of the issue of electronic personhood. Altogether, this book presents a welcome contribution to the research community.

Umeå University
Umeå, Sweden

Virginia Dignum

Acknowledgments

This book is the result of a doctoral seminar on "Algorithmic Governance and Governance of Algorithms" which took place at the University of Tartu on 23 May 2019. The seminar was organised as part of the IT Law Programme of the University of Tartu, the University of Tartu School of Law and the University of Tartu Doctoral School of Economics and Innovation, supported by the European Union, European Regional Development Fund (University of Tartu's ASTRA project PER ASPERA), in co-operation with the Robotics & AI Law Society (RAILS).

The editors are grateful to the people who made this doctoral seminar possible and would like to thank the Director of the University of Tartu School of Law, Professor Gaabriel Tavits, the Director of the PhD programme, Associate Professor Merike Ristikivi, the Director of the IT Law programme, Anette Aav, as well as Katri Holst and Kertu Kirjasto for their administrative support. In addition, the editors would especially like to acknowledge Irene Negri for her excellent assistance in the editing of this book.

This book has been supported by the Estonian Research Council's grant no. PRG124 and by the Research Project "Machine Learning and AI-Powered Public Service Delivery", RITA1/02-96-04, funded by the Estonian Government.

Contents

Algorithmic Governance and Governance of Algorithms: An Introduction

Marta Cantero Gamito and Martin Ebers

Contents

Abstract The use of algorithms is more than ever replacing human decision-making. Naturally, this raises concerns about how to govern AI-powered technologies. This chapter introduces the potential as well as the threat(s) posed by decision-making by algorithms (algorithmic governance) and provides an up-to-date overview of the state of art and the existing legislative initiatives in this field (governance of algorithms).

Sections 1–3 written by Martin Ebers; Sections 4 and 5 written by Marta Cantero Gamito.

M. Cantero Gamito
CUNEF, Law Department, Madrid, Spain

University of Tartu, School of Law, Tartu, Estonia

M. Ebers (✉)
Humboldt University of Berlin, Law Faculty, Berlin, Germany

University of Tartu, School of Law, Tartu, Estonia
e-mail: martin.ebers@ut.ee

M. Ebers, M. Cantero Gamito (eds.), *Algorithmic Governance and Governance of Algorithms*, Data Science, Machine Intelligence, and Law 1,
https://doi.org/10.1007/978-3-030-50559-2_1

1 The Promise and Perils of Algorithm-Based Technologies

Algorithm-based technologies[1] such as artificial intelligence (AI)[2] and smart robotics[3] are increasingly pervading our lives. They are deployed in many sectors ranging from retail and production, finance and transport to healthcare and security.[4] They come in different forms and shapes, as personal assistants on our smartphones, search engines, translation apps, data-mining programs, scoring systems, medical diagnosis systems, price algorithms, expert trading systems, and in physical manifestations, such as self-driving cars, drones, unmanned underwater vehicles, surgical robots, personal robots and social robots.

Many of these systems have the potential to improve our lives as well as the overall economic and societal welfare. AI-powered systems can lead to better healthcare services,[5] safer and cleaner transport systems,[6] better working conditions,[7] higher productivity[8] and new innovative products, services and supply chains.[9] They can also benefit the public sector in a number of ways;[10] for example, by automating repetitive and time-consuming tasks, or by providing public agencies with more accurate and detailed information, forecasts and predictions, which in turn might lead to personalized public services tailored to individual circumstances. AI-powered systems may even help to respond to key global challenges, such as the climate change[11] and the novel coronavirus pandemic.[12]

However, as with every disruptive technology, AI and smart robotics come not only with benefits but also with substantial risks, raising a broad variety of ethical

[1]Generally speaking, an algorithm can be understood as "sets of defined steps structured to process instructions/data to produce an output"; Kitchin (2017).

[2]Popular definitions of AI are equally unrefined. For different definitions cf. Samoili et al. (2020), High Level Expert Group on AI (AI HLEG) (2019a), McCarthy (2007), Russell and Norvig (2011), pp. 1 et seq.

[3]There is currently no generally accepted definition of the term "robot" either. The AI HLEG describes robotics as "AI in action in the physical world", but points out that robots cannot be equated with AI systems, since robotics also uses technologies that lie outside of artificial intelligence; cf. AI HLEG (2019a), p. 4.

[4]For an overview on different use-cases cf. OECD (2019a), pp. 47 et seq.; International Electrotechnical Commission (IEC) (2018), pp. 45 et seq.

[5]Abu-Nasser (2017) and Gray (2018).

[6]Anderson et al. (2016), pp. 9 et seq.; OECD (2019a), pp. 48 et seq.

[7]Arntz et al. (2016) and OECD (2019b).

[8]Autor and Salomons (2018).

[9]Charalambous et al. (2019) and World Economic Forum (2017).

[10]Freeman Engstrom et al. (2020).

[11]Vinuesa et al. (2020).

[12]Kritikos, European Parliamentary Research Service (EPRS) (2020), pp. 1–2; Dumbrava, European Parliamentary Research Service (EPRS) (2020).

and legal challenges.[13] Algorithmic systems can unpredictably harm people's life, health, and property. They can also affect fundamental values on which western societies are founded, leading to breaches of fundamental rights, including the rights to human dignity and self-determination, privacy and personal data protection, freedom of expression and of assembly, non-discrimination, or the right to an effective judicial remedy and a fair trial, as well as consumer protection.[14]

Automated or algorithmic decision making (ADM) systems are particularly of concern. Many important decisions, which were historically made by people, are now either made by machines or at least prepared by them.[15] We live in a "scored society"[16] in which citizens, consumers and legal entities are increasingly subject to actions and decisions taken by or with the assistance of AI systems. AI increases the possibilities to track and score the daily habits of people. Companies from various industries collect, analyze, acquire, share, trade, and utilize data on billions of people in order to discern patterns, predict the likely behavior of people through scoring systems, and act accordingly. As a result, there is not only the danger that AI systems—in violation of data protection law—are used by state authorities or private companies for *mass surveillance*. Rather, the widespread use of algorithms for preparing or even making decisions is also criticized on the grounds of *discrimination.* A number of examples show that ADM procedures are by no means neutral but can perpetuate and even exacerbate biases in various ways. Compared to human decision-making, ADM systems pose a particular challenge: while it is true that human decision-making is not immune to mistakes and biases, algorithmic decisions can have a much larger effect, as the software not only decides dozens or hundreds of cases, but rather tens of thousands or more.

AI systems can also unintentionally or intentionally lead to *manipulation*. Social media platforms and search engines use AI systems to channel, prioritize and filter information—with potentially detrimental effect on the right to freedom of information, the right to freedom of expression, media pluralism, and the political discourse in general.[17] Moreover, the insights gained by AI powered systems can be used by companies or political parties to exploit or trigger irrational behavior—a practice which in the end led to the well-known *Cambridge Analytica* scandal.[18]

Particularly troublesome is the fact that not only private companies, but also *governmental institutions* are increasingly relying on algorithmic systems to analyze and predict behavior in order to make decisions. Tax offices have started using

[13]For an in-depth analysis cf. Ebers (2020). According to the Stanford AI Index 2019, the ethical challenges most mentioned across 59 ethical AI framework documents were: fairness; interpretability and explainability; transparency, accountability; data privacy, reliability, robustness and security; Perrault et al. (2019), p. 149.

[14]Cf. Council of Europe, Committee of Experts on Internet Intermediaries (MSI-NET) (2017), Raso et al. (2018).

[15]Cf. AlgorithmWatch (2019).

[16]Keats and Pasquale (2014).

[17]Pariser (2011), Sunstein (2017) and Epstein (2014).

[18]Cf. Nix (2016); moreover Rubinstein (2014).

algorithms to predict abuse and fraud in tax returns and to allocate cases for human review.[19] In social welfare systems, algorithms are used to determine whether a citizen should be flagged because of an increased risk of irregularities, or potential fraud.[20] In the field of public security, many agencies use AI systems to detect terrorists,[21] to screen people at the border,[22] and to predict and respond to crime (predictive policing).[23] In the US, algorithmic prognosis instruments are even used by courts to calculate the likelihood of an accused person committing another crime while on parole.[24] In China, the government has implemented a Social Credit System which is intended to standardize the assessment of citizens' and businesses' economic and social reputations.[25]

The risks associated with AI systems and smart robotics are further intensified and amplified by the specific characteristics of many of these technologies, including opacity (black box-effect),[26] complexity, unpredictability and partially autonomous behavior—all of which may make it hard to verify compliance with the existing legal rules. Due to these characteristics, both enforcement authorities and affected persons might lack the means to verify how a given algorithmic decision was taken and whether the relevant rules were respected.

[19]DeBarr and Harwood (2004).

[20]For Austria see https://algorithmwatch.org/en/story/austrias-employment-agency-ams-rolls-out-discriminatory-algorithm/; for Finland, https://www.tieto.com/en/success-stories/2018/the-city-of-espoo-a-unique-experiment/; for the Netherlands https://bijvoorbaatverdacht.nl/; for Spain https://algorithmwatch.org/en/story/spain-legal-fight-over-an-algorithms-code/; for Sweden https://algorithmwatch.org/en/rogue-algorithm-in-sweden-stops-welfare-payments/. Accessed 29 Apr 2020.

[21]In the EU, the European Commission is funding the DANTE experiment, an anti-terrorism project (Detecting and analysing terrorist-related online contents and financing activities), aimed at using automated decision-making against terrorism, https://www.h2020-dante.eu/. Accessed 29 Apr 2020.

[22]In the EU, many countries use "iBorderCtrl", a system tested in Hungary, Greece and Latvia to screen non-EU nationals at EU borders, using automated interviews with a virtual border guard, based on "deception detection" technology; https://www.iborderctrl.eu/The-project. Accessed 29 Apr 2020.

[23]Barrett (2017), Ferguson (2012), p. 317; Rich (2016) and Saunders et al. (2016).

[24]Such processes are used at least once during the course of criminal proceedings in almost every US state; Barry-Jester et al. (2015). More than 60 predictive tools are available on the market, many of which are supplied by companies, including the widely-used COMPAS system from Northpointe.

[25]Hvistendahl (2017), Botsman (2017). Chen et al. (2018), pointing out that the Social Credit System has not—at least for now—employed AI technologies, real-time data or automated decisions, despite foreign media reports to the contrary.

[26]The notion of black-box AI refers to such scenarios, where we can see only input data and output data for algorithm-based systems without understanding exactly what happens in between. Burrell (2016), Leese (2014), Mittelstadt et al. (2016), p. 6; Pasquale (2015).

2 From Algorithmic Governance to Governance of Algorithms

2.1 The Current Legal Landscape

There is currently not a single country (or supranational organization) in the world with legislation that explicitly takes into account the problematic characteristics of algorithmic systems in general.

Apart from a few exceptions, there are also no special rules for AI systems and smart robotics in particular. Admittedly, special regulation exists for self-driving vehicles,[27] drones,[28] and high-frequency trading.[29] Moreover, in the European Union, the General Data Protection Regulation (GDPR) contains rudimentary provisions for fully automated decisions.[30] In addition, the EU Consumer Rights Directive 2011/83/EU, as amended by the "New Deal for Consumers", includes an obligation to provide information on the use of automatically generated personalised prices.[31] Moreover, the so-called P2B (Platform-to-business) Regulation 2019/1150 requires providers of online search engines to "set out the main parameters, which individually or collectively are most significant in determining ranking and the relative importance of those main parameters, by providing an easily and publicly

[27]In the US, most of the states have either enacted legislation or executive orders governing self-driving vehicles; cf. National Conference of State Legislatures, Autonomous Vehicles State Bill Tracking Database, http://www.ncsl.org/research/transportation/autonomous-vehicles-legislative-database.aspx. Accessed 29 Apr 2020. In 2017, the House of Representatives passed a bill for a "Self Drive Act" which was supposed to lay out a basic federal framework for autonomous vehicle regulation but, ultimately, failed to be considered on the Senate floor. For the EU, see Expert Group on Liability and New Technologies—New Technologies Formation (2019).

[28]In the EU, the Regulation on Civil Aviation 2018/1139 addresses issues of registration, certification, and general rules of conduct for operators of drones—however, without regulating civil liability directly; cf. Bertolini (2018).

[29]See esp. Art. 17, Art. 48(6) MiFID II (Directive 2014/65/EU on markets in financial instruments) and Commission delegated Regulation (EU) 2017/589 of 19 July 2016 supplementing Directive 2014/65/EU of the European Parliament and of the Council with regard to regulatory technical standards specifying the organizational requirements of investment firms engaged in algorithmic trading, OJ 31 March 2017 L 87/417.

[30]Regulation (EU) 2016/679 of the European Parliament and of the Council of 27 April 2016 on the protection of natural persons with regard to the processing of personal data and on the free movement of such data, and repealing Directive 95/46/EC, OJ L 119, 4 May 2016. Art. 22 GDPR prohibits fully automated decisions; for those decisions, Art. 13 (2) lit. f and Art. 14 (2) lit. g GDPR establish moreover a special obligation for data controllers to provide information. However, these provisions have a rather limited scope of application, because they only apply to decisions "based solely on automated processing" of data. Since most algorithmically prepared decisions still involve a human being, the majority of ADM procedures is not covered by the aforementioned provisions; cf. Martini (2019), pp. 10 et seq.

[31]According to Art. 6(1) (ea) Consumer Rights Directive 2011/83/EU as amended by Directive 2019/2161/EU, the trader may have to inform the consumer "that the price has been personalised on the basis of an automated decision making process".

available description".[32] Furthermore, some countries have issued rules for automated algorithm-based administrative decisions, such as Canada with its "Directive on Automated Decision-Making".[33] In France, the Digital Republic Act (Loi no. 2016-1321 du 7 octobre 2016 pour une République numérique), provides that, in the case of state actors taking a decision "on the basis of algorithms", individuals have a right to be informed about the "principal characteristics" of the decision-making system.[34]

However, all these rules deal only with individual aspects without providing a satisfactory answer to the problems described above. Certainly, many countries and sometimes also international and supranational organizations have laws, norms and rules that are relevant for AI and robotics—ranging from constitutional principles (rule of law, democracy),[35] human rights,[36] and (international) humanitarian law;[37] to administrative and criminal law protecting inter alia fair procedures;[38] to special laws that could help to mitigate the described problems such as data protection law, cybersecurity law, product safety and product liability law, competition law, consumer law; and many other fields. These laws, however, were not made with AI and smart robotics in mind.

Accordingly, it is difficult to gauge to what extent existing legislation sufficiently regulates the undesirable implications of AI.

2.2 *Existing Initiatives to Regulate AI and Robotics in a Nutshell*

Since the beginning of 2017, many governments in the world have begun to develop national strategies for the promotion, development, and use of AI systems. Still, as *Tim Dutton*—a Canadian Senior Policy Advisor who regularly updates a summary of different AI policies—observes, no two strategies are alike.[39] Instead, national (and international) initiatives focus on a wide variety of aspects such as research and development programs, skills and education, data and digital infrastructure,

---

[32] Art. 5(2) Regulation 2019/1150 on promoting fairness and transparency for business users of online intermediation services (P2B Regulation), OJ L 186, 11 July 2019.

[33] Government of Canada, 'Directive on Automated Decision-Making', https://www.tbs-sct.gc.ca/pol/doc-eng.aspx?id=32592. Accessed 29 Apr 2020.

[34] For more details see Edwards and Veale (2018).

[35] Cf. for example Council of Europe, European Commission for the Efficiency of Justice (CEPEJ) (2018).

[36] Cf. Council of Europe, Committee of Experts on Internet Intermediaries (MSI-NET) (2017); Raso et al. (2018).

[37] Margulies (2018).

[38] On AI and administrative law cf. Oswald and Grace (2016), Cobbe (2018) and Coglianese and Lehr (2017).

[39] Dutton (2018). Cf. also the overview by Thomas (2018).

technical standardization, AI-enhanced public services, ethics and inclusion, and sometimes also legal standards. Whereas some countries have laid down specific and comprehensive AI strategies (e.g. China, the UK, France), some are integrating AI technologies within national technology or digital roadmaps (e.g. Denmark, Australia), while still others have focused on developing a national AI R&D strategy (US).[40]

In the US, most notably, the government already relied heavily under the Obama administration on the liberal notion of the free market.[41] In its report "Preparing for the Future of Artificial Intelligence", published in October 2016,[42] the White House Office of Science and Technology Policy (OSTP) explicitly refrains from a broad regulation of AI research and practice. Instead, the report highlights that the government should aim to fit AI into existing regulatory schemes, suggesting that many of the ethical issues related to AI can be addressed through increasing transparency and self-regulatory partnerships.[43] The Trump administration, too, sees its role not in regulating AI and robotics but in "facilitating AI R&D, promoting the trust of the American people in the development and deployment of AI-related technologies, training a workforce capable of using AI in their occupations, and protecting the American AI technology base from attempted acquisition by strategic competitors and adversarial nations"—thus maintaining US American leadership.[44] In January 2020, the White House published a draft memorandum outlining ten principles which federal agencies should consider when devising laws and rules for the use of AI in the private sector, but stressed—again—that a key concern was limiting regulatory "overreach".[45]

By contrast, the European Union focusses in its AI strategy (published in April 2018) and its Whitepaper on AI (published in February 2020) not only on the potential impact of AI on competitiveness but also on its social and ethical implications, underpinning that compliance with European ethical norms, legal requirements and social values is essential to create "an ecosystem of trust".

Beyond the European Union, several international organizations have also taken the initiative to reflect on the future legal framework for AI and robotics, such as the Organisation for Economic Cooperation and Development (OECD) with its principles on AI, adopted in May 2019,[46] and the new AI Policy Observatory which aims

[40]Delponte (2018), p. 22.

[41]For a detailed discussion of the various AI strategies in the US, the EU, and the UK, see Cath et al. (2018).

[42]Executive Office of the [US] President—National Science and Technology Council Committee on Technology (2016). The report followed five workshops and a public request for Information, cf. Executive Office of the [US] President—National Science and Technology Council Committee on Technology (2016), p. 12.

[43]Executive Office of the [US] President—National Science and Technology Council Committee on Technology (2016).

[44]Trump (2019). Cf. also Shepardson (2018).

[45]Office of Management and Budget (OMB), the White House (2019).

[46]OECD (2019c).

to help policymakers to implement the AI principles;[47] the United Nations (UN) with its several activities on AI;[48] and the Council of Europe with its "European Ethical Charter on the Use of Artificial Intelligence in Judicial Systems and their environment", adopted at the end of 2018,[49] and its ad hoc committee on AI (CAHAI) with the specific task to examine the possibility of creating a legal framework for the development, design and application of artificial intelligence, based on Council of Europe's standards on human rights, democracy and the rule of law.[50]

Apart from these initiatives, hundreds of different ethical AI guidelines have emerged during the past years. The number of ethical guidelines developed by governments, tech companies, and industry has grown exponentially in recent years—so much that researchers had to develop special tools to provide an overview.[51] However as laudable as this work may be, it should be clear that soft law as such will not suffice. The work on ethical principles and guidelines can lay the groundwork for subsequent legislation, providing orientation on the possible content of legal rules. However, the main problem is that ethical guidelines and self-regulatory initiatives by industries are non-binding.[52] In addition, these principles are often too abstract to provide sufficiently detailed guidance. As it has been pointed out, "[m]uch of the debate about ethics seems increasingly focused on companies avoiding regulation. Unable or unwilling to properly provide regulatory solutions, ethics is seen as the 'easy' or 'soft' option which can help structure and give meaning to existing self-regulatory initiatives."[53] Indeed, ethical guidelines and self-regulation should not be used as an escape from (hard) regulation.

3 The European Union's AI Strategy

3.1 The EU as the Global Regulatory Standard-Setter

The regulatory efforts of the EU are particularly noteworthy. The EU is widely acknowledged to be the regulatory standard-setter,[54] especially in the field of data protection and its respective flagship, the General Data Protection Regulation.[55] Although it has been argued that the GDPR puts European companies at a competitive disadvantage compared with firms in countries such as China and the US by

---

[47]https://oecd.ai/. Accessed 29 Apr 2020.

[48]https://www.itu.int/pub/S-GEN-UNACT-2018-1. Accessed 29 Apr 2020.

[49]Council of Europe, European Commission for the Efficiency of Justice (CEPEJ) (2018).

[50]www.coe.int/cahai. Accessed 29 Apr 2020.

[51]See Fjeld et al. (2020).

[52]Saurwein et al. (2015).

[53]Wagner (2018), pp. 108 et seq.

[54]Bradford (2020). Cremona and Scott (2019).

[55]Smuha (2019).

imposing restrictions on the collection and use of personal data, which is especially relevant in the context of AI systems,[56] this criticism did not alter the fact that the GDPR has been one of Europe's most successful regulatory export products. Since its entry into force in May 2018, the GDPR has had a significant impact on data protection policy and enforcement beyond the EU in many jurisdictions, including California and Brazil.[57]

With regard to AI, the European Union also seems to be willing to take the lead in developing regulatory standards. In 2019, the new president of the European Commission, *Ursula von der Leyen*, announced that in her first hundred days in office, she "would put forward legislation for a coordinated European approach on the human and ethical implications of Artificial Intelligence",[58] thereby building on existing protective regulation as well as on the work of the High Level Expert Group on AI that published AI Ethics Guidelines in April 2019. In the same vein, German Chancellor *Angela Merkel* announced in June 2019 at the G20 summit in Japan, that "it will be the job of the next Commission to deliver something so that we have regulation similar to the General Data Protection Regulation that makes it clear that artificial intelligence serves humanity".[59]

Clearly, the European Union has not been the only regulatory power to claim a leading role in the governance of AI and robotics. However, no other country has taken a stance as strong as the EU's promise to bring forth new binding legislation in this field. Against this background, the following observations focus solely on the political efforts of the EU to regulate algorithms, AI and robotics, while initiatives from other states and international organizations are not taken into account.

3.2 *The European Parliament's Resolution of February 2017*

In the European Union, it was above all the European Parliament (EP) that first developed a strategy for an EU-wide regulation of AI and robotics. In February 2017, the EP passed a Resolution "with recommendations to the Commission on Civil Law Rules on Robotics".[60] The resolution called for the creation of a "European Agency for Robotics and AI" consisting of regulators and external technical and ethical experts who could provide the "technical, ethical and regulatory

---

[56]On the frictions between Big Data Practices based on AI and the GDPR see Zarsky (2017) and Humerick (2018).

[57]For a review of 24 jurisdictions cf. https://theword.iuslaboris.com/hrlaw/whats-new/the-impact-of-the-gdpr-outside-the-eu. Accessed 29 Apr 2020.

[58]Von der Leyen (2019), p. 13.

[59]Kayali (2019).

[60]European Parliament, 'Resolution of 16 February 2017 with recommendations to the Commission on Civil Law Rules on Robotics, P8_TA(2017)0051'. The resolution does not include unembodied AI. Instead, AI is understood as an underlying component of "smart autonomous robots". Critically, Cath et al. (2018).

expertise needed to support the relevant public actors, at both Union and Member State level, in their effort to ensure a timely, ethical and well-informed response to the new opportunities and challenges",[61] and could monitor robotics-based applications, identify standards for best practice and, where appropriate, recommend regulatory measures, define new principles, and address potential consumer protection issues.[62] Moreover, the Resolution recommended introducing an EU-wide registration system for specific categories of advanced robots.[63]

Apart from that, the EP proposed to develop a Charter on robotics consisting of a code of ethical conduct for researchers and designers to "act responsibly and with absolute consideration for the need to respect the dignity, privacy and safety of humans".[64] In addition, the EP asked the European Commission to clarify the liability of industry and autonomous robots when harm or damages occur and to adopt new rules on liability if necessary.[65]

3.3 The European Commission's AI Strategy and the Work of the High-Level Expert Group on AI

On 25 April 2018, the European Commission published its Communication "Artificial Intelligence for Europe".[66] The document—complemented by another communication of 7 December 2018[67]—outlined three pillars as the core of the proposed strategy: (i) boosting the EU's technological and industrial capacity and AI uptake across the economy, (ii) preparing for socio-economic changes brought by AI, and (iii) ensuring an appropriate ethical and legal framework based on the Union's values in line with the Charter of Fundamental Rights of the EU.

To support its implementation, the Commission established the "High-Level Expert Group on Artificial Intelligence"[68] (AI HLEG) and mandated it with the drafting of two documents in particular: (i) AI Ethics Guidelines that build on the work of the European Group on Ethics in Science and New Technologies[69] and of the European Union Agency for Fundamental Rights,[70] and (ii) Policy and

[61] European Parliament, Resolution, No. 16.

[62] European Parliament, Resolution, No. 17.

[63] European Parliament, Resolution, No. 2.

[64] European Parliament, Resolution, p. 19.

[65] European Parliament, Resolution, Nos. 49 et seq.

[66] European Commission (2018a).

[67] European Commission (2018b).

[68] https://ec.europa.eu/digital-single-market/en/high-level-group-artificial-intelligence. Accessed 29 Apr 2020.

[69] European Group on Ethics in Science and New Technologies (2018).

[70] The European Union Agency for Fundamental Rights (FRA), an independent EU body funded by the EU budget, started a new project on 'Artificial Intelligence, Big Data and Fundamental Rights'

Investment Recommendations. At the same time, the European AI Alliance,[71] an open multi-stakeholder platform with over 2700 members, was set up to provide broader input for the work of the AI HLEG.

At the end of 2018, the AI HLEG presented its first draft, "Ethics Guidelines for Trustworthy AI".[72] After an open consultation which generated feedback from more than 500 contributors, the AI HLEG published the final version at the beginning of April 2019,[73] followed in June 2019 by the "Policy and Investment Recommendations for Trustworthy AI".[74] The ethics guidelines are neither an official document from the European Commission nor legally binding. They are also not intended as a substitute for any form of policy-making or regulation, nor to deter from the creation thereof.[75] Rather, the ethics guidelines should serve as a "living document" and starting point for the discussion about "Trustworthy AI for Europe".[76]

One of the main goals of these guidelines is to ensure that the development and use of AI follows a human-centric approach, according to which AI is not seen as a means in itself but as a tool to enhance human welfare and freedom. To this end, the AI HLEG propagates "trustworthy AI" which is (i) lawful, complying with all applicable laws and regulations; (ii) ethical, ensuring adherence to ethical principles and values; and (iii) robust, both from a technical and social perspective. The document aims to offer guidance on achieving Trustworthy AI by setting out in Chapter I fundamental rights and ethical principles AI should comply with. From those fundamental rights and principles, Chapter II derives seven key requirements (human agency and oversight; technical robustness and safety; privacy and data governance; transparency; diversity, non-discrimination, and fairness; societal and environmental wellbeing; and accountability), which then lead in Chapter III to a concrete but non-exhaustive assessment list to apply the requirements of Chapter II, offering AI practitioners guidance. During a piloting phase, organizations had the possibility to test the assessment list for trustworthy AI artificial intelligence.[77]

The work of the AI HLEG is accompanied by evaluations of the current EU safety and liability framework. To this end, the Commission set up the "Expert Group on Liability and New Technologies",[78] which published at the end of 2019 its findings on how liability regime should be designed—and, where necessary changed—in

in 2018 with the aim of helping create guidelines and recommendations in these fields. Cf. https://fra.europa.eu/en/about-fra/introducing-fra. Accessed 29 Apr 2020.

[71] https://ec.europa.eu/digital-single-market/en/european-ai-alliance. Accessed 29 Apr 2020.

[72] AI HLEG (2018).

[73] AI HLEG (2019b). Moreover, AI HLEG (2019a) published the document 'A Definition of AI: Main Capabilities and Disciplines'.

[74] AI HLEG (2019c).

[75] AI HLEG (2019b), p. 3.

[76] AI HLEG (2019b), p. 3.

[77] European Commission (2019), p. 7.

[78] European Commission (2018a), p. 16.

order to rise to the challenges emerging digital technologies bring with them.[79] Another paper of early 2020, prepared by the Joint Research Center (JRC) of the European Commission, deals with the topic "Robustness and Explainability of Artificial Intelligence".[80]

3.4 The European Commission's White Paper on AI

On the basis of this preparatory work, the Commission published on 19 February 2020 its long-awaited White Paper "On Artificial Intelligence - A European approach to excellence and trust" to launch a public consultation on the future legal framework for AI.[81] The White Paper on AI is supplemented by two other documents, namely a report on the safety and liability implications of new technologies,[82] and the European data strategy.[83] In addition, the Commission announced in its current work program that binding legislation on AI will be proposed in Q4 2020, in particular regarding safety, liability, fundamental rights and data aspects.[84]

The White Paper highlights, first of all, possible adjustments to the existing EU legislative framework to AI.[85] The specific characteristics of many AI systems—such as the already mentioned black box effect, complexity, unpredictability and semi-autonomous behaviour—could lead to new safety risks and liability gaps and could make effective enforcement of EU legislation more difficult, as algorithmic decisions cannot be traced and therefore cannot be checked for legal compliance.[86] The European Commission is therefore considering an evaluation, and if necessary, adjustment of the current legal framework with regard to "the protection of fundamental rights and consumer rights" and with regard to "existing EU product safety and liability legislation".[87]

In addition to possible adjustments, the White Paper considers that *new legislation specifically on AI* may be needed "in order to make the EU legal framework fit for the current and anticipated technological and commercial developments". Taking into consideration potential future regulation of AI, the White Paper takes a *risk-based approach*: as any regulatory intervention should be targeted and proportionate, the Commission does not want to introduce new mandatory legislation for all AI

[79] Expert Group on Liability and New Technologies—New Technologies Formation (2019).
[80] Hamon et al. (2020).
[81] European Commission (2020d).
[82] European Commission (2020c).
[83] European Commission (2020b).
[84] European Commission (2020a), Annex I.
[85] European Commission (2020d), pp. 13 et seq.
[86] European Commission (2020d), p. 12.
[87] European Commission (2020d), p. 13.

systems, but only for so-called "high-risk AI systems", whereas "low-risk AI systems" should be subject to a voluntary labelling scheme.[88]

For high-risk cases, the White Paper identifies the following requirements that could be included in the regulatory framework:[89]

- Requirements regarding the *quality of training data* to guarantee representative and comprehensive data-sets in compliance with privacy and data protection rules
- *Record and data keeping requirements* to enable a regulatory review and enforcement by allowing AI decisions to be traced back and verified
- *Adequate information and transparency requirements* regarding the use, limitations and capabilities of high-risk AI systems
- *Requirements on robustness and accuracy* of high-risk AI systems by avoiding and minimizing the different risks it may be vulnerable to
- *Human involvement*, ranging from human oversight before a decision is implemented, to the possibility of human intervention in real time or afterwards
- *Specific requirements for remote biometric identifications* such as *facial recognition.*

In order to ensure compliance of high-risk AI applications with these mandatory requirements, the European Commission envisages the creation of prior conformity assessments, which could include procedures for testing, inspections or certification—and "checks of the algorithms and of the data sets used in the development phase".[90]

Should the risk-based approach favoured by the Commission prevail, it is essential from the point of view of legal certainty, that the criteria for "high risk AI applications" are clearly defined. In this respect, however, the Commission's statements remain rather vague. On the one hand, the Commission is of the opinion that AI applications should be considered high-risk only if both the sector *and* the intended use involve significant risks.[91] In this regard, the Commission plans to exhaustively list these high-risk sectors, as for instance healthcare, transport, energy and parts of the public sector. On the other hand, the Commission acknowledges that the use of AI may in exceptional circumstances be considered as a high-risk by itself, irrespective of the sector concerned and where the above mentioned requirements would still apply. As examples, the White Paper mentions the use of AI for recruitment purposes and remote biometric identification, as well as "specific applications affecting consumer rights".[92] However, it remains open which AI applications the Commission has specifically in mind.

For the future discussion, it will therefore be important first to identify the shortcomings of the current (national and European) legal framework for the use

[88]European Commission (2020d), p. 17.

[89]European Commission (2020d), pp. 18 et seq.

[90]European Commission (2020d), p. 23.

[91]European Commission (2020d), p. 17.

[92]European Commission (2020d), p. 18.

of AI systems, and to determine the corresponding regulatory requirements for specific risks. For the time being, as the White Paper signals, both high-risk and non-high-risk existing AI applications remain entirely subject to already existing rules. Accordingly, the chapters of this volume provide an examination from different legal backgrounds, enriching the legal debate on the governance dimension of AI technologies.

4 Overview of This Book

In view of the different and fast-evolving developments of AI technologies manifold questions emerge. Some of these questions have been discussed above. As noted, the most pressing issue concerns the power of algorithms. There is no doubt that the expansion of AI is changing our world. The emergence of this new technological reality allows algorithms to make or influence decisions (*algorithmic governance*). Most of the online services we use nowadays rely on algorithms. Not only individuals and corporations, but also public authorities are delegating decision-making to algorithms, resulting in potential threats to privacy, non-discrimination, consumer protection, freedom of expression or the right to a fair trial among others. Certainly, this phenomenon is impacting how law is designed, applied and interpreted. The massive diffusion of algorithms is indeed fueling debates on how to govern this power adequately; there is an unrelenting urge to address the *governance of algorithms*.

This book brings those two dimensions together. The different contributions discuss the impact of algorithms and the different possibilities of forthcoming regulation. This volume combines some of the key topics and the challenges posed by algorithms, namely transparency and due process in automated decision-making, liability, privacy, consumer protection and copyright.

Niklas Eder focuses on the increasing use of algorithms for individual profiling and its implications for privacy, non-discrimination and equal treatment. In his chapter, the key question revolves around how to address the challenges of profiling without individually undermining the fundamental rights involved. Eder argues that, in finding solutions, scholars have disregarded the boundaries between different categories of fundamental rights, which may ultimately undermine the relevance and essence of these rights. The chapter critically engages in the reasons why we, as lawyers, should care about establishing a proper definition and delimitation of the fundamental rights concerned; a proper definition and delineation of fundamental rights will significantly shape the normative choices to be made. Eder also provides a useful discussion on the existing trade-off between fundamental rights and the actual value-added of services relying on machine learning based data analysis innovations. The chapter concludes by suggesting a narrow interpretation of the potential fundamental rights affected by profiling, as the means to fully exploit the potential benefits of machine learning algorithms.

Francesca Palmiotto examines the consequences of the increasing use of algorithms for the purposes of collecting evidence and decision-making in criminal proceedings. More often than not, algorithms used for such purposes are not disclosed, making it very difficult to assess whether the process transforming the input into output respects fair trial rights. Specifically, Palmiotto focuses on the possibility for these algorithms to contain miscodes, and how these miscodes, which can be technical, scientific or legal, remain invisible as a result of algorithmic opacity. She uses the Harm Assessment Risk Tool (HART)—a software used by the UK police for supporting custody decisions and for assessing the risk of future offending by arrestees—as a use case, and shows how potential miscodes may remain hidden as a result of the opacity of the algorithms involved in the software. She discusses how the identified miscodes may impact fair trial rights and due process, such as preventing the contestability of the decision. Palmiotto concludes that the use of algorithms in criminal proceedings should be aligned with existing procedural guarantees. The risk is otherwise a dysfunctional system of criminal justice where fundamental rights are impaired.

In their chapter, *Seili Suder* and *Merle Erikson* explore the consequences of a dystopian phenomenon that is closer to reality than what we might think, microchipping employees. This practice is added to a wide array of mechanisms for controlling labour, namely computer screen recording, video surveillance, keystroke monitoring, location tracking and social media monitoring, inter alia. The pervasive nature of microchipping, most notably constant surveillance, outweighs the possible benefits associated with its use. The risks of escalating beyond merely monitoring employee's activities poses significant threats to privacy. Hence, the chapter carefully examines the privacy implications of microchipping in light of GDPR. The authors conclude that the collection of employee information triggers the application of the GDPR and that the employee's consent is not a sufficient legal ground for the processing of personal data. As a result, the authors require the existence of a legitimate interest of the employer. However, they argue, this alone is not enough. Suder and Erikson propose the creation of specific national rules protecting microchipped employees, esp. by prohibiting the compulsory use of microchipping, which should be restricted only to those occupations where monitoring employee's activities is critical for the job.

Tomás García-Micó offers an extensive overview of issues related to the liability of autonomous devices. Using the example of surgical robots, the chapter explores different implications of granting electronic personhood to autonomous goods. The questions to be addressed in this regard have sparked a scholarly debate on the capacity of existing civil rules to address the problems posed by robotics. For the most part, the current EU product safety framework already covers some of the risks posed by autonomous devices. Yet, for the purposes of legal certainty, concrete rules are still required to govern new emerging realities. García-Micó provides a careful analysis of current proposals, such as the European Parliament Resolution containing recommendations to the Commission on Civil Law Rules on Robotics also discussed above. The chapter concludes that granting legal personhood to autonomous devices, as suggested by the European Parliament Resolution, is unnecessary and inadequate. Instead, it recommends strengthening other existing measures such as a mandatory insurance scheme for producers, which would facilitate compensation for persons having suffered harm caused by these devices.

Previous chapters have already shown that algorithms generally do not operate in an unregulated environment. In most cases, existing frameworks can accommodate the challenges posed by their increasing use. However, such accommodation requires a proper scrutiny and a serious assessment to identify whether new rules are needed. *Federico Galli* provides a detailed evaluation of the capacity of GDPR and the Unfair Commercial Practices Directive (UCPD) to properly cope with the problems arising from behavioral advertising; e.g. manipulation. This granular analysis reveals that the UCPD appears to be the most suitable instrument to address behavioral advertising. The article concludes with a call for a re-examination of the consumer policy approach embedded in the UCPD, and points at the insights offered by behavioral economics for the much-needed revision of the standard 'average consumer' in the twenty-first century.

Jasper Siems explores the suitability of the legal solutions offered by intellectual property (IP) rules to protect artificial neural networks (ANN). ANN are critical for deep learning. The chapter illustrates how existing IP regimes, such as copyright and patent protection are inadequate to protect ANN. As explained in the chapter, copyright and patent protection are generally anthropocentric. This largely rules out the possibility of protecting the intellectual creation of ANN by copyright or patent laws. Siems turns to trade secret protection rules as an alternative for the protection of the works by neural networks. The result is inconclusive; trade secret rules would have to be refined to accommodate ANN. It is argued however, that the application of trade secret rules is, despite its limitations, preferred to a rushed and potentially underdeveloped instrument that would inadequately address data ownership.

The last chapter deals with the legal status of computer-generated works (CGWs). *Ying Ye* and *Mike Adcock* provide an analysis of the Chinese copyright framework to explore how copyrightability and authorship operate in the case of CGWs. The chapter offers interesting illustrations of the possibilities of AI technologies to develop creative output, so far exclusively attributed to humans. Examples include the Xiaoice application, a poem generator, or Alibaba's Luban, an AI-based graphic design tool. The analysis focuses on the application of copyright rules through the examination of the scope of Chinese copyright law to cover CGWs. Unlike the UK copyright law that explicitly defines CGWs, Chinese legislation, similarly to other countries, has not yet incorporated this development. By proposing a separation between authorship and ownership, Ye and Adcock conclude that Chinese copyright law may extend protection to CGWs under existing rules which allow "work for hire" and the "work of legal person or entity".

5 The Way Forward

The continuous development and increasing use of AI technologies has created significant governance challenges. However, the legal and political implications of the use of algorithms for decision-making go beyond governance questions, triggering a social debate on the meaning and reach of algorithms in our future life. A proper prognosis should consider the challenges of scalability as well as the potential benefits of AI-powered systems for business and society. In this light, any (serious) legislative initiative should consider the possibilities and limitations of providing a legal framework sufficiently flexible to fit the purpose. From a legal perspective, potential threats to fundamental rights triggered by the pervasive use of algorithms are particularly pressing.

While a ‘wait-and-see’ approach is suitable for fast-pace emerging technologies, a clear regulatory roadmap such as the one provided in the European Commission’s latest White Paper is a sensible starting point. The significance of the issues involved, i.e. potential fundamental rights violations, favors a short-term strategy. Therefore, the proposed EU regulatory plan based on a risk-based approach, if executed appropriately, would ensure a much-needed democratic scrutiny and legal certainty for the ongoing development of AI. Moreover, the EU’s strategy and political vision on AI is set to influence an international discussion. The proposed EU framework will most likely have extraterritorial implications, similarly to the GDPR and other sectors.[93] The Commission attempts to assert a distinctive European position on AI regulation for a trustworthy ecosystem. Two issues would require a closer look here.

First, the European Commission aims at establishing a framework for excellence and trust in which it is “paramount” that the requirements are applicable to all relevant economic operators providing AI-enabled products or services in the EU, “regardless of whether they are established in the EU or not”.[94] This means that some non-EU operators will be subject to the abovementioned *prior conformity assessment* to verify that AI is trustworthy, secure and in line with European values and rules. This is important with regard to fundamental rights protection. The Commission clearly points at risks to fundamental rights and specifically mentions the need to ensure effective judicial redress for parties negatively affected by AI.[95] In order to monitor and facilitate compliance, the White Paper mentions the role of standards. Standards might also contribute to specify future mandatory requirements.[96] In this regard, it will be important to examine which values underpin the standard-setting process.

Secondly, a closer look should be paid to the coordination of global compliance. Regulation is also currently being drafted outside of the EU. In January 2020, the US

[93]Cf. Cremona and Scott (2019) and Bradford (2020).

[94]European Commission (2020d), p. 22.

[95]European Commission (2020d), p. 23.

[96]European Commission (2020d), p. 18.

has published a set of principles which US agencies must meet when drafting AI regulations.[97] This initiative is complemented by the proposal for an Algorithmic Accountability Act of 2019,[98] currently under discussion. Given that many AI applications are built through a complex supply chain, it is expected that any legislative initiative will most likely have a global impact.

In conclusion, while fueling debates around the world, the EU Commission's White Paper is certainly welcome and can be considered a sensible path towards the development of legislation for AI as well as to guide the global development of AI applications. However, the EU approach should not be rushed into a regulatory race. The concrete proposals expected towards late 2020 will provide some apparent regulatory relief, but the legislator cannot, and should not, overlook the fact that AI is constantly evolving. The expected evolution of AI will require a flexible regulatory approach that could make the EU framework a long-lasting one.

The rapid technological breakthroughs and the complexity of the sector make it particularly challenging for scholars and policymakers, who seek to understand and evaluate this phenomenon and its multifaceted implications. This volume contains a thoughtful analysis of some of the key legal topics that arise with regard to AI, framing the discussions questions and challenges to be faced in upcoming days.

References

Abu-Nasser B (2017) Medical expert systems survey. Int J Eng Inf Syst 7:218

AlgorithmWatch (2019) Automating society. Taking stock of automated decision-making in the EU. https://algorithmwatch.org/wp-content/uploads/2019/01/Automating_Society_Report_2019.pdf. Accessed 29 Apr 2020

Anderson JM et al (2016) Autonomous vehicle technology. A guide for policymakers. RAND, Santa Monica California. https://www.rand.org/content/dam/rand/pubs/research_reports/RR400/RR443-2/RAND_RR443-2.pdf. Accessed 29 Apr 2020

Arntz M et al (2016) The risk of automation for jobs in OECD countries: a comparative analysis. OECD Social, Employment and Migration Working Papers, No. 189. OECD Publishing, Paris. https://doi.org/10.1787/5jlz9h56dvq7-en. Accessed 29 Apr 2020

Autor D, Salomons A (2018) Is automation labor-displacing? Productivity growth, employment, and the labor share. NBER Working Paper, No. 24871. https://doi.org/10.3386/w24871. Accessed 29 Apr 2020

Barrett L (2017) Reasonably suspicious algorithms: predictive policing at the United States border. N Y Univ Rev Law Soc Change 41(3):327

Barry-Jester AM et al (2015) The new science of sentencing. The Marshall Project, 4 Apr 2015. https://www.themarshallproject.org/2015/08/04/the-new-science-of-sentencing#.xXEp6R5rD. Accessed 29 Apr 2020

[97]Office of Management and Budget (OMB), the White House (2019).

[98]www.wyden.senate.gov/imo/media/doc/Algorithmic%20Accountability%20Act%20of%202019%20Bill%20Text.pdf. Accessed 29 Apr 2020.

Bertolini A (2018) Artificial intelligence and civil law: liability rules for drones. Study commissioned by the European Parliament's Policy Department for citizens' rights and constitutional affairs at the request of the JURI Committee, PE 608.848
Botsman R (2017) Big data meets Big Brother as China moves to rate its citizens. Wired UK, 21 Oct 2017. https://www.wired.co.uk/article/chinese-government-social-credit-score-privacy-invasion. Accessed 29 Apr 2020
Bradford A (2020) The Brussels effect: how the European Union rules the world. Oxford University Press, Oxford
Burrell J (2016) How the machine "thinks": understanding opacity in machine learning algorithms. Big Data Soc, January-June, 1
Cath C et al (2018) Artificial intelligence and the "Good Society": the US, EU, and UK approach. Sci Eng Ethics 24(2):505
Charalambous E et al (2019) AI in production: a game changer for manufacturers with heavy assets. https://www.mckinsey.com/business-functions/mckinsey-analytics/our-insights/ai-in-production-a-game-changer-for-manufacturers-with-heavy-assets. Accessed 29 Apr 2020
Chen Y et al (2018) "Rule of Trust": the power and perils of China's social credit megaproject. Columbia J Asian Law 32(1):1. https://ssrn.com/abstract=3294776. Accessed 29 Apr 2020
Cobbe J (2018) Administrative law and the machines of government: judicial review of automated public-sector decision-making. A pre-review version of a paper in Legal Studies, Forthcoming. https://ssrn.com/abstract=3226913. Accessed 29 Apr 2020
Coglianese C, Lehr D (2017) Regulating by robot: administrative decision making in the machine-learning era. Georgetown Law J 105:1147. https://ssrn.com/abstract=2928293. Accessed 29 Apr 2020
Council of Europe, Committee of Experts on Internet Intermediaries (MSI-NET) (2017) Algorithms and human rights - Study on the human rights dimensions of automated data processing techniques and possible regulatory implications. DGI(2017)12. https://rm.coe.int/study-hr-dimension-of-automated-data-processing-incl-algorithms/168075b94a. Accessed 29 Apr 2020
Council of Europe, European Commission for the Efficiency of Justice (CEPEJ) (2018) European Ethical Charter on the use of artificial intelligence in judicial systems and their environment. Strasbourg, 3-4 Dec 2018. https://rm.coe.int/ethical-charter-en-for-publication-4-december-2018/16808f699c. Accessed 29 Apr 2020
Cremona M, Scott J (2019) EU law beyond EU borders. Oxford University Press, Oxford
DeBarr D, Harwood M (2004) Relational mining for compliance risk, presented at the Internal Revenue Service Research Conference. http://www.irs.gov/pub/irs-soi/04debarr.pdf [https://perma.cc/Y9F8-RWNK]. Accessed 29 Apr 2020
Delponte L (2018) European Artificial Intelligence (AI) leadership, the path for an integrated vision, Study requested by the ITRE Committee of the European Parliament. PE 626.074. http://www.europarl.europa.eu/RegData/etudes/STUD/2018/626074/IPOL_STU(2018)626074_EN.pdf. Accessed 29 Apr 2020
Dumbrava C, European Parliamentary Research Service (EPRS) (2020) Tracking mobile devices to fight coronavirus. PE 649.384. http://www.europarl.europa.eu/RegData/etudes/BRIE/2020/649384/EPRS_BRI(2020)649384_EN.pdf. Accessed 29 Apr 2020
Dutton T (2018) An overview of national AI strategies. Medium, 28 June 2018. https://medium.com/politics-ai/an-overview-of-national-ai-strategies-2a70ec6edfd. Accessed 29 Apr 2020
Ebers M (2020) Regulating AI and robotics: ethical and legal challenges. In: Ebers M, Navas Navarro S (eds) Algorithms and law. Cambridge University Press, Cambridge, pp. 37–99
Edwards L, Veale M (2018) Enslaving the algorithm: from a "right to an explanation" to a "right to better decisions"? IEEE Security & Privacy, May/June 46
Epstein R (2014) How Google could end democracy. U.S. News & World Report, 9 June 2014. https://www.usnews.com/opinion/articles/2014/06/09/how-googles-search-rankings-could-manipulate-elections-and-end-democracy. Accessed 29 Apr 2020
European Commission (2018a) Communication 'Artificial Intelligence for Europe'. COM(2018) 237 final

European Commission (2018b) Communication 'Coordinated plan on Artificial Intelligence'. COM (2018) 795 final
European Commission (2019) Communication 'Building trust in human-centric Artificial Intelligence'. COM(2019) 168 final
European Commission (2020a) Communication 'Commission Work Programme 2020. A Union that strives for more'. COM(2020) 37 final
European Commission (2020b) Communication 'European strategy for data'. COM(2020) 66 final
European Commission (2020c) Report on the safety and liability implications of artificial intelligence, the Internet of Things and robotics. COM(2020) 64 final
European Commission (2020d) White Paper 'On Artificial Intelligence – A European approach to excellence and trust'. COM(2020) 65 final
European Group on Ethics in Science and New Technologies (2018) Statement on artificial intelligence, robotics and "autonomous" systems. Brussels, 9 Mar 2018. https://ec.europa.eu/info/news/ethics-artificial-intelligence-statement-ege-released-2018-apr-24_en. Accessed 29 Apr 2020
Executive Office of the [US] President - National Science and Technology Council Committee on Technology (2016) Preparing for the future of artificial intelligence (OSTP report). https://obamawhitehouse.archives.gov/sites/default/files/whitehouse_files/microsites/ostp/NSTC/preparing_for_the_future_of_ai.pdf. Accessed 29 Apr 2020
Expert Group on Liability and New Technologies – New Technologies Formation (2019) Liability for artificial intelligence and other emerging technologies. https://doi.org/10.2838/573689. https://ec.europa.eu/newsroom/dae/document.cfm?doc_id=63199. Accessed 29 Apr 2020
Ferguson AG (2012) Predictive policing and reasonable suspicion. Emory Law J 62:259
Fjeld J et al (2020) Principled artificial intelligence: mapping consensus in ethical and rights-based approaches to principles for AI. Berkman Klein Center Research Publication No. 2020-1, 15 Jan 2020. https://doi.org/10.2139/ssrn.3518482. Accessed 29 Apr 2020
Freeman Engstrom D et al (2020) Government by algorithm: artificial intelligence in federal administrative agencies, Report submitted to the administrative conference of the United States. https://www-cdn.law.stanford.edu/wp-content/uploads/2020/02/ACUS-AI-Report.pdf. Accessed 29 Apr 2020
Gray A (2018) 7 amazing ways artificial intelligence is used in healthcare. World Economic Forum, 20 Sep 2018. www.weforum.org/agenda/2018/09/7-amazing-ways-artificial-intelligence-is-used-in-healthcare. Accessed 29 Apr 2020
Hamon R et al (2020) Robustness and explainability of artificial intelligence – From technical to policy solutions. Publications Office of the European Union, Luxembourg. https://doi.org/10.2760/57493 (online), JRC119336. Accessed 29 Apr 2020
High Level Expert Group on AI (AI HLEG) (2018) Draft "Ethics Guidelines for trustworthy AI", Working Document for stakeholders' consultation. Brussels, 18 Dec 2018. https://ec.europa.eu/digital-single-market/en/news/draft-ethics-guidelines-trustworthy-ai. Accessed 29 Apr 2020
High Level Expert Group on AI (AI HLEG) (2019a) A definition of AI: main capabilities and disciplines. Brussels, 9 Apr 2019. https://ec.europa.eu/newsroom/dae/document.cfm?doc_id=56341. Accessed 29 Apr 2020
High Level Expert Group on AI (AI HLEG) (2019b) Ethics Guidelines for trustworthy AI (Ethics Guidelines). Brussels, 8 Apr 2019. https://ec.europa.eu/newsroom/dae/document.cfm?doc_id=58477https://ec.europa.eu/futurium/en/ai-alliance-consultation. Accessed 29 Apr 2020
High Level Expert Group on AI (AI HLEG) (2019c) Policy and investment recommendations for trustworthy AI. Brussels, 26 June 2019. ec.europa.eu/newsroom/dae/document.cfm?doc_id=60343. Accessed 29 Apr 2020
Humerick M (2018) Taking AI personally: how the E.U. must learn to balance the interests of personal data privacy & artificial intelligence. Santa Clara High Technol Law J 34:393
Hvistendahl M (2017) Inside China's vast new experiment in social ranking. Wired, 14 Dec 2017. https://www.wired.com/story/age-of-social-credit/. Accessed 29 Apr 2020

International Electrotechnical Commission (IEC) (2018) White Paper 'Artificial Intelligence across industries'. https://basecamp.iec.ch/download/iec-white-paper-artificial-intelligence-across-industries-en/. Accessed 29 Apr 2020
Kayali L (2019) Next European Commission takes aim at AI. Artificial intelligence will be the next GDPR. Politico, 18 July 2019. https://www.politico.eu/article/ai-data-regulator-rules-next-european-commission-takes-aim/. Accessed 29 Apr 2020
Keats D, Pasquale FA (2014) The scored society: due process for automated predictions. Washington Law Rev 89:1
Kitchin R (2017) Thinking critically about and researching algorithms. Inf Commun Soc 20 (1):1–14
Kritikos M, European Parliamentary Research Service (EPRS) (2020) Ten technologies to fight coronavirus. PE 641.543
Leese M (2014) The new profiling: algorithms, black boxes, and the failure of anti-discriminatory safeguards in the European Union. Secur Dialog 45(5):494
Margulies P (2018) The other side of autonomous weapons: using artificial intelligence to enhance IHL compliance. Roger Williams University Legal Studies Paper No. 182, 12 June 2018. https://ssrn.com/abstract=3194713. Accessed 29 Apr 2020
Martini M (2019) Fundamentals of a regulatory system for algorithm-based processes. Speyer. https://www.vzbv.de/sites/default/files/downloads/2019/07/19/martini_regulatory_system_algorithm_based_processes.pdf. Accessed 29 Apr 2020
McCarthy J (2007) What is artificial intelligence? www-formal.stanford.edu/jmc/whatisai/. Accessed 29 Apr 2020
Mittelstadt BD et al (2016) The ethics of algorithms: mapping the debate. Big Data Soc, July-September, 1
Nix A (2016) Speech at the 2016 Concordia annual summit in New York. https://www.youtube.com/watch?v=n8Dd5aVXLCc. Accessed 29 Apr 2020
OECD (2019a) Artificial intelligence in society. OECD Publishing, Paris. https://doi.org/10.1787/eedfee77-en. Accessed 29 Apr 2020
OECD (2019b) Going digital: shaping policies, improving lives. OECD Publishing, Paris. https://doi.org/10.1787/9789264312012-en. Accessed 29 Apr 2020
OECD (2019c) Recommendation of the Council on Artificial Intelligence. OECD/LEGAL/0449. legalinstruments.oecd.org/en/instruments/OECD-LEGAL-0449. Accessed 29 Apr 2020
Office of Management and Budget (OMB), The White House (2019) Draft Memorandum for the Heads of executive departments and agencies, Guidance for regulation of artificial intelligence applications. www.whitehouse.gov/wp-content/uploads/2020/01/Draft-OMB-Memo-on-Regulation-of-AI-1-7-19.pdf. Accessed 29 Apr 2020
Oswald M, Grace J (2016) Intelligence, policing and the use of algorithmic analysis: a freedom of information-based study. J Inf Rights Policy Pract 1(1). https://journals.winchesteruniversitypress.org/index.php/jirpp/article/view/16. Accessed 29 Apr 2020
Pariser E (2011) The filter bubble: what the internet is hiding from you. Penguin Press, New York
Pasquale F (2015) The black box society: the secret algorithms that control money and information. Harvard University Press, London
Perrault R et al (2019) The AI Index 2019 Annual Report. AI Index Steering Committee, Human-Centered AI Institute. Stanford University, Stanford, Dec 2019. https://hai.stanford.edu/sites/g/files/sbiybj10986/f/ai_index_2019_report.pdf. Accessed 29 Apr 2020
Raso F et al (2018) Artificial intelligence & human rights: opportunities & risks. Berkman Klein Center Research Publication, No. 2018-6, 25 Sep 2018. https://doi.org/10.2139/ssrn.3259344. Accessed 29 Apr 2020
Rich M (2016) Machine learning, automated suspicion algorithms, and the fourth amendment. Univ Pa Law Rev 164:871
Rubinstein I (2014) Voter privacy in the age of big data. Wisconsin Law Rev:861
Russell S, Norvig P (2011) Artificial intelligence: a modern approach. 3rd ed. Pearson Education, London

Samoili S et al (2020) AI Watch. Defining artificial intelligence. Towards an operational definition and taxonomy of artificial intelligence. EUR 30117 EN. Publications Office of the European Union, Luxembourg. http://publications.jrc.ec.europa.eu/repository/bitstream/JRC118163/jrc118163_ai_watch._defining_artificial_intelligence_1.pdf. Accessed 29 Apr 2020
Saunders J et al (2016) Predictions put into practice: a quasi-experimental evaluation of Chicago's predictive policing pilot. J Exp Criminol 12:347
Saurwein F et al (2015) Governance of algorithms: options and limitations. Info 17(6):35
Shepardson D (2018) Trump administration will allow AI to "freely develop" in U.S.: official. Technology News, 10 May 2018. https://www.reuters.com/article/us-usa-artificialintelligence/trump-administration-will-allow-ai-to-freely-develop-in-u-s-official-idUSKBN1IB30F. Accessed 29 Apr 2020
Smuha NA (2019) From a "race to AI" to a "race to AI regulation": regulatory competition for artificial intelligence. https://doi.org/10.2139/ssrn.3501410. Accessed 29 Apr 2020
Sunstein CR (2017) #Republic. Divided democracy in the age of social media. Princeton University Press, Princeton
Thomas V (2018) Report on artificial intelligence: Part I - the existing regulatory landscape. Howtoregulate.com, 14 May 2018. https://www.howtoregulate.org/artificial_intelligence/. Accessed 29 Apr 2020
Trump D (2019) Executive order on maintaining American leadership in artificial intelligence. 11 Feb 2019. https://www.whitehouse.gov/presidential-actions/executive-order-maintaining-american-leadership-artificial-intelligence/. Accessed 29 Apr 2020
Vinuesa R et al (2020) The role of artificial intelligence in achieving the sustainable development goals. Nat Commun 11:233. https://doi.org/10.1038/s41467-019-14108-y. Accessed 29 Apr 2020
Von der Leyen U (2019) Political Guidelines for the next European Commission 2019-2024, A Union that strives for more: my agenda for Europe. 19 July 2019. https://ec.europa.eu/commission/sites/beta-political/files/political-guidelines-next-commission_en.pdf. Accessed 29 Apr 2020
Wagner B (2018) Ethics as an escape from regulation: from ethics-washing to ethics-shopping? In: Hildebrandt (ed) Being profiled: *cogitas ergo sum*. Amsterdam University Press, Amsterdam, pp. 84–89
World Economic Forum (2017) Impact of the fourth industrial revolution on supply chains. October 2017. http://www3.weforum.org/docs/WEF_Impact_of_the_Fourth_Industrial_Revolution_on_Supply_Chains_.pdf. Accessed 29 Apr 2020
Zarsky TZ (2017) Incompatible: the GDPR in the age of big data. Seton Hall Law Rev 47(4):995

Privacy, Non-Discrimination and Equal Treatment: Developing a Fundamental Rights Response to Behavioural Profiling

Niklas Eder

Contents

Abstract In the diverse attempts to identify fundamental rights implications of behavioural profiling, the lines between the right to privacy, non-discrimination and equal treatment have been blurred. Scholars have struggled to develop coherent approaches to the widespread practice of evaluating and differentiating between individuals on the basis of correlative relations between random, causally unrelated categories in large data sets. This chapter suggests a response to these practices and establishes clear boundaries between the rights. It is argued that the right to

N. Eder (✉)
Yale Law School, New Haven, CT, USA

European Law School, Humboldt University of Berlin, Berlin, Germany

University of Leipzig, Leipzig, Germany
e-mail: niklas.eder@uni-leipzig.de

M. Ebers, M. Cantero Gamito (eds.), *Algorithmic Governance and Governance of Algorithms*, Data Science, Machine Intelligence, and Law 1,
https://doi.org/10.1007/978-3-030-50559-2_2

non-discrimination should be interpreted narrowly, thus not applying to large parts of behavioural profiling. Extending its scope to random categories would jeopardise the distinguished quality of the right to effectively prohibit the most appalling and morally reprehensible differentiations. The scope of the right to privacy, conversely, has an open-ended structure and can be further evolved in order to respond to novel threats. However, it is first and foremost the right to equal treatment, which carries great, even though little acknowledged potential to provide a normative framework for behavioural profiling. The engagement with it can encourage, frame and respond to a much needed societal debate on machine learning based data analysis. Difficulties which occur in the application of the right to equal treatment are manifestations of a larger societal challenge. The shared question, which both, fundamental rights lawyers and society at large have to answer is how to accommodate the new way of generating knowledge and differentiating between individuals within our conventional way to understand the world, to reason and to differentiate.

1 Introduction

Accounts, in which scientists, politicians and journalists describe the revolutionary potential of artificial intelligence (AI) are countless.[1] One of the fields in which the revolutionary impact of machine learning (ML) algorithms is no longer potential and fictitious but present and real is the analysis of data relating to human behaviour, also called behavioural profiling.[2] Behavioural profiling has mainly been developed in advertising in order to better target potential clients. However, today, behavioural profiling is used in a great variety of fields and has effects on many spheres of life. Popular examples include, but are not limited to sectors, such as finance, health and security.[3]

With view to fundamental rights, scholars are pondering about the impact of as well as the appropriate response to behavioural profiling. In the diverse attempts to identify its fundamental rights implications, the lines between the right to privacy, non-discrimination and equal treatment have been blurred. This chapter re-establishes clear boundaries between the rights and suggests a way to develop a normative framework for behavioural profiling. It proceeds in three steps: The first part describes the processes underlying behavioural profiling and points out the challenges which it poses to fundamental rights research (2). The second part describes the scope of the rights as well as the objectives and values which they pursue (3). The third part identifies challenges in the application of the rights and suggests starting points to develop responses (4). In these three steps, the article

[1]For a sober and expansive analysis of the functioning and impact of 'Big Data', see Mayer-Schönberger and Cukier (2013).

[2]For a detailed description and critical analysis see Zuboff (2019).

[3]See for example Wei et al. (2015); on algorithms deciding the likelihood of recidivism, see Corbett-Davies et al. (2016).

answers the following questions: If individuals are evaluated and treated on the basis of correlative relations between random categories, detected by machine learning algorithms in the analysis of large data sets, why should we care? What fundamental rights are affected by practices of behavioural profiling? Where does the application of fundamental rights pose challenges, and how can we solve these challenges? The EU Charter of Fundamental Rights (CFR) and the European Convention of Human Rights (ECHR) serve as models for reflections, which can be transferred to similar fundamental rights established elsewhere.

I argue that the scope of the right to non-discrimination should be interpreted narrowly, thus not addressing large parts of behavioural profiling. There are many non-discrimination issues in machine learning based data analysis. Differentiating individuals on the basis of correlations between random categories is not one of them. The right to non-discrimination is a distinguished right, which does not address all kinds of but only the most appalling and morally reprehensible differentiations. Extending its scope to behavioural profiling would further the protection of all individuals in all spheres of life on the cost of the distinguished protection of vulnerable groups and of sensitive spheres of life. An extensive interpretation would water down the legal and moral force of right to non-discrimination and is therefore counterproductive. The scope of the right to privacy, conversely, has an open-ended structure and can be further evolved in order to respond to novel threats deriving from behavioural profiling. Notable scholarly controversy does not regard the question if, but how to evolve to right to privacy. I suggest to refrain from evolving the values of the right, structurally transforming it from an individual to a group right. Approaches that suggest recognising the identity of groups as a value worth protecting risk to dilute the foundations of the right.

The main argument of this paper centres on the right to equal treatment. It departs from the observation that behavioural profiling can be regarded as a technology whose very purpose it is to differentiate between individuals. Its broad-scale use brings with it a new paradigm of differentiating between individuals. The right to equal treatment, correspondingly, has the purpose of assessing differentiations and is, therefore, an adequate starting point to develop a fundamental rights response. The application of the right to equal treatment to behavioural profiling, while providing a promising starting point, also poses great challenges. Machine learning algorithms pose great challenges to the application of the right to equal treatment, as the normative foundations, on the basis of which they compare and differentiate, are antithetical to the ones which humans intuitively and courts conventionally built on. The conventional approach compares situations on the basis of causal relations within the subject matter. Machine learning algorithms, contrarily, compare situations in consideration of all available information and decide on the relevancy of differences on the basis of correlation. The hard question that fundamental rights scholars are confronted with in face of behavioural profiling is to decide if and how to accommodate the new way in which machine learning algorithms compare and differentiate within the conventional assessment of the right to equal treatment.

2 Understanding the Challenges

To set the foundations for examining current fundamental rights challenges, I briefly elaborate on the notion of behavioural profiling and its underlying technological process (Sect. 2.1). Subsequently, I point out where, in the broad debate on fundamental rights implications of behavioural profiling, distinguishing between fundamental rights poses challenges (Sect. 2.2).

2.1 Behavioural Profiling

Profiling describes a practice of personalising the treatment of individuals. Behavioural profiling can relate to different types of profiling.[4] For the course of this article, a broad understanding is taken as a basis. Profiling can be described as behavioural when either the data which is analysed (input), or the information that it generates (output) relates to the behaviour of humans.[5] However, neither the input nor the output has to refer exclusively to behaviour. The input includes data relating to activities which appear rather random, such as clicking behaviour, mouse moving patterns or the speed with which a person scrolls through his or her social media account as well as data on activities whose relevance is rather obvious, such as restaurant visits, movements patterns and financial transactions. Often, the input will also include information which relates not to behaviour, but to circumstances in which a person finds him- or herself: such as his or her employment status, professional network or living situation. The output, too, includes information relating to behaviour, such as shopping and travel preferences, as well as information about circumstances, including information regarding a person's financial resources, health or social environment. The qualification of profiling as behavioural, hence, puts the main focus on information relating to behaviour but does not exclude other types of information.

The technological process of machine learning based profiling can be approached with two notions: groups and correlations.[6] The grouping part of this process is, when simplified, fairly straight forward: algorithms group individuals on the basis of their input data.[7] The categories on the basis of which individuals are grouped are open-ended. Individuals are clustered to the same group when they have similar data

---

[4]For a variety of perspectives on profiling see Hildebrandt and Gutwirth (2008); for an overview on fields, in which profiling is used, see Wachter and Mittelstadt (2018).

[5]For the distinction between data and information this article relies on the differentiation of Mayer-Schönberger and Cukier (2013).

[6]The brief account here cannot be achieved without major simplifications. For a more detailed account see Hildebrandt and Koops (2010).

[7]For a detailed account on what kind of "groups" machine learning algorithms create, see Kammourieh et al. (2017).

points in one category. A group, in this context, is not to a stable and enclosed entity. Rather, there are a great number of groups which partly overlap and whose composition permanently changes in dependence of the behaviour of individuals. Also, every individual is part of a large number of groups, as every data point clusters him or her with different individuals.

To generate profiles and infer information about individuals which has previously been unknown, machine learning algorithms identify correlations between groups and outcomes. The generation of profiles happens in two steps. In the first step, the algorithm is trained in large data sets and generates knowledge in the form of statistical correlations between input data and outcomes.[8] The data sets used in training can be based entirely on anonymous data, placing the generation of knowledge outside the General Data Protection Regulation (GDPR).[9] In the second step, the generated knowledge is used to evaluate individual cases.[10] Data relating to an individual is matched with the groups in which the algorithm has identified to be relevant predictors of certain outcomes. On the basis of the previously identified correlations and patterns, the algorithm calculates how likely certain outcomes are in an individual case.[11] The analysis of an individual case begins with rather harmless input data, such as clicking behaviour, mouse moving patterns or scrolling speed, and ends with an output which can be extremely sensitive or economically decisive, such as sexual orientation, psychological conditions or financial capacity.[12] Inferring previously unknown information and generating predictions which can be used as a basis for evaluations of individuals is the very purpose of the machine learning based analysis.[13] This information is consequently used to personalise the way individuals are encountered.

2.2 *Difficulties in Distinguishing Between the Fundamental Rights*

Research on fundamental rights implications of behavioural profiling is broad and diverse.[14] Before suggesting a way to differentiate between the right to privacy, non-discrimination and equal treatment, I will demonstrate in what parts of the

[8]For an early, but convincing account see Hildebrandt (2006).

[9]For detailed accounts on the legal framework of profiling under the GDPR see Wachter et al. (2017b, 2018) and Wachter and Mittelstadt (2018).

[10]For a brief but precise explanation on the process of generating knowledge, see Hildebrandt (2006).

[11]Kammourieh et al. (2017).

[12]Wachter and Mittelstadt (2018).

[13]Mayer-Schönberger and Cukier (2013).

[14]For overviews see Hildebrandt and Koops (2010). For overviews see Mittelstadt et al. (2016), Gonçalves (2017).

research a clear differentiation is hard to accomplish but crucial to have. Among the large scholarly engagement, two main fields of concern can be identified. While these fields are related, each is characterised by a particular focus of interest.

The first field departs from privacy concerns provoked by the increased generation of information about individuals and groups.[15] This research is firstly triggered by the observation that it is no longer the collection of data, but the generation of knowledge through the analysis of large data sets which exposes extensive and often intimate information about individuals.[16] Secondly, scholars have pointed out that it is no longer individuals, but groups who are in the main focus of data analysis.[17] Against the background of these two observations, scholars have reassessed the efficiency of current laws and rights and have criticised that they fail to provide sufficient protection. In response, scholars, on the one hand, have called for new standards with which machine learning based data analysis should have to comply, such as requirements of explainability,[18] transparency[19] and reasonableness.[20] On the other hand, scholars have tried to push beyond the "anthropocentric focus" of current laws and rights.[21] As it is groups who are in the main focus of machine learning based data analysis, it is argued, privacy and data protection rights have to apply to groups, too.[22] It is this latter call for "Group Privacy", which blurs the line between the right to privacy and non-discrimination.

Research in the second field finds its basis in anti-discrimination law and examines how states and, in liberal democracies more importantly, companies change the way they encounter individuals on the basis of information generated through machine learning algorithms.[23] Scholars in this field have faced two challenges. The first challenge lies in the discovery of hidden discriminations.[24] Only rarely will machine learning algorithms differentiate directly on the basis of protected categories: Under the GDPR, data relating to protected categories is "sensitive" and its use is prohibited (Art. 9 GDPR). Anyhow, data sets often contain information which correlates with protected categories and algorithms discriminate on the basis of proxies. Furthermore, critical scholars examine how biased structures manifest

[15]For overviews on AI's implications on privacy see Schermer (2011), Rubinstein (2013); for contributions on Group Privacy, see Taylor et al. (2017a).

[16]On the legal status of the generated information see Wachter and Mittelstadt (2018).

[17]Mantelero (2016) and Kammourieh et al. (2017).

[18]Wachter et al. (2017b).

[19]Wachter et al. (2017a).

[20]Wachter and Mittelstadt (2018).

[21]Floridi (2017).

[22]Among the versatile scholarly engagement see for example Floridi (2014), Kammourieh et al. (2017), Mittelstadt (2017); for an overview see Taylor et al. (2017b), Van der Sloot (2017).

[23]See for example: Leese (2014), Roberts (2015), Barocas and Selbst (2016), Kim (2017) and Hacker (2018).

[24]Also discussed under the notion indirect discrimination, on this term see Schabas (2015).

themselves in machine learning algorithms.[25] They expose how structural biases in data sets, in the development process of algorithms as well as in their deployment, translate into biased results in behavioural profiling. It is indeed an important and difficult task to detect hidden and structural biases. However, the challenge which scholars face here does not lie in the distinction between different fundamental rights: it is clearly the right to non-discrimination, which is relevant. It is in the second challenge in non-discrimination research on machine learning data analysis, where the differentiation between different fundamental rights is difficult but crucial. The second challenge begins, where the first ends: it arises when there is neither an apparent nor a hidden link between the categories on the basis of which machine learning based algorithms differentiate and categories which are listed or have been recognised to be protected under the right to non-discrimination. As groups which are relevant in behavioural profiling are structurally different from groups conventionally protected by the right to non-discrimination, the question which arises, is, if and how the scope of the right should be extended to these groups. Does, and if so under what conditions, the right to non-discrimination apply to groups formed on the basis of correlations between random categories? In the attempt to answer these questions, the demarcation of the right to non-discrimination from other fundamental rights poses challenges. Leaving the framework of well-known fundamental rights, scholars have focused on the notion of "algorithmic fairness".[26] Building on rights in data-protection and anti-discrimination law,[27] scholars here, too, have blurred the lines of conventional distinctions.

While blurring lines can be fruitful and necessary in face of new technologies, conceptual clarity remains an important precondition to penetrate behavioural profiling's normative implications successfully. The rights to privacy, non-discrimination and equal treatment remain important pillars in the construction of the much-required normative framework. In the following, to provide the basis for an allocation of the challenges within these fundamental rights, I establish their scope as well as their normative foundation.

3 Privacy, Non-Discrimination and Equal Treatment

Difficulties regarding the assessment of fundamental rights implications of behavioural profiling can be located on two levels. The first relates to the scope of application, the second to the normative foundations of the rights. In face of behavioural profiling, it is not only unclear whether the scopes of the rights apply. There is also controversy about what the objectives and values are, which each right protects and whether these objectives and values are affected by behavioural

[25] Noble (2018).

[26] See for example: Hacker and Wiedemann (2017) and Zarsky (2016).

[27] Hacker (2018).

profiling. I engage with the right to privacy (Sect. 3.1), before looking at the right to non-discrimination and equal treatment (Sect. 3.2). The CFR and the ECHR serve as an example for my description of the rights, which can be transferred to similar fundamental rights established elsewhere.

3.1 *The Right to Privacy*

The right to privacy entails a larger number of guarantees (Sect. 3.1.1) and serves the realisation of values, which are fundamental within liberal democracies (Sect. 3.2.2).

3.1.1 Scope of the Right

Art. 7 CFR[28] as well as Art. 8(1) ECHR[29] establish the right to private life (here also: right to privacy) in almost the same wording.[30] In addition to that, Art. 8 CFR establishes the right to data protection, which the ECJ regards as a specific manifestation of the right to privacy and, at least in grave cases, assess together with Art. 7 CFR.[31] According to Art. 52(3) and Art. 53 CFR, the CFR is interpreted in accordance with the ECHR, and the CFR shall provide at least the same if not more extensive protection than the ECHR. In its application of the CFR, the European Court of Justice (ECJ) gives the interpretations of the European Court of Human Rights (ECtHR) of corresponding rights in the ECHR great relevance[32] and it is the ECtHR's case law which has shaped the concrete guarantees deriving from the protection of the right to privacy.[33]

The ECtHR's concept of privacy encompasses a large number of guarantees.[34] The right establishes not only negative but also positive obligations of states to actively protect privacy.[35] According to the ECtHR, it is neither possible nor necessary to attempt an exhaustive definition of the notion "private life" or the scope of the right.[36] Rather, the ECtHR has recognised various guarantees which it regards necessary to protect privacy. Besides guarantees relating to the physical,

[28]"Everyone has the right to respect for his or her private and family life, home and communications."

[29]"Everyone has the right to respect for his private and family life, his home and his correspondence."

[30]Before the CFR entered into force, the ECJ has ruled that the right to the respect of private life derives from the constitutional traditions of the member states and is therefore protect by the European legal order: ECJ 5 October 1994, X v Commission (C-404/92 P).

[31]ECJ 8 April 2014, Digital Rights Ireland and Seitlinger a.o. (C-293/12 and C-594/12).

[32]Kokott and Sobotta (2015).

[33]For an overview see: Jacobs et al. (1996).

[34]For an overview on co-existing definitions, see: Jacobs et al. (1996).

[35]ECtHR 13 June 1979, Marckx/Belgium (6833/74); Jacobs et al. (1996).

[36]ECtHR 16 December 1992, Niemietz/Germany (13710/88) para 29.

psychological and moral integrity,[37] the right to privacy includes protections of one's own image,[38] personal reputation[39] and personal information[40] as well as the protection against state surveillance.[41] Additionally, the right to privacy includes guarantees such as the right to choose one's gender identity,[42] discover one's origin[43] and live a life according to one's ethnic identity.[44] The generation of information about individuals and groups, as well as the use of that information to differentiate between them, is not explicitly addressed by one of the existing guarantees. The hard question, whether machine learning data analysis falls into the scope of the right to privacy, can only be decided when taking into account the purpose of the right and its underlying values.

3.1.2 Objectives and Values

The objectives which are pursued with the right to privacy are numerous, too.[45] The ECtHR explains that the right to privacy, besides the rather straightforward objectives of the physical and psychological integrity of a person, serves the protection of an individual's identity, of personal development as well as free self-determination.[46] Also, in order to protect free self-determination, the ECtHR considers personal autonomy an important principle guiding its interpretation of the right.[47] The diversity of objectives which the ECtHR has recognised is mirrored in the old and diverse scholarly debate on the values underlying the right to privacy.

Despite all controversy, it is persuading to recognise that the right to privacy finds its basis in both, liberty as well as human dignity. Scholars have argued that the American understanding of privacy is shaped by the underlying value of liberty, while the European understanding is shaped by the underlying value of dignity.[48] Limiting the European conception of privacy to the value of dignity, however, stands in contrast to the broad interpretation of the right by the ECtHR. The expressively

[37]ECtHR 26 March 1985, X and Y/the Netherlands (8978/80) para 22; for the psychological integrity see for example: ECtHR 6 February 2001, Bensaid/the United Kingdom (44599/98).

[38]See for example: ECtHR 24 June 2004, Von Hannover/Germany (59320/00).

[39]See for example: ECtHR 7 February 2012, Axel Springer AG/Germany (39954/08).

[40]On that the decision of the ECJ: ECJ 8 April 2014, Digital Rights Ireland and Seitlinger a.o. (C-293/12 and C-594/12).

[41]ECtHR 21 June 2011, Shimovolos/Russia (30194/09).

[42]See for example: ECtHR 16 July 2014, Hämäläinen/Finland (37359/09).

[43]See for example: ECtHR 13 February 2003, Odièvre/France (42326/98).

[44]See for example: ECtHR 18 January 2001, Chapman/the United Kingdom (27238/95).

[45]On the debate whether privacy, at all, has intrinsic value, see: Solove (2002); for an overview of protected interests see: Harris (2009).

[46]ECtHR 29 April 2002, Pretty/the United Kingdom (2346/02) para 61.

[47]ECtHR 29 April 2002, Pretty/the United Kingdom (2346/02).

[48]Whitman (2004).

stated objectives, as well as the particular guarantees established in the case law of the ECtHR find their basis in both, liberty as well as dignity. Guarantees such as freedom from state surveillance, or to live a life according to one's choosing and ethnic identity, serve the protection of liberty. Other guarantees are more closely related with human dignity, such as the guarantees to have control over one's own image, personal reputation and to discover one's origin. Taking further into account the objectives of the right to privacy as acknowledged by the ECtHR, much speaks in favour of giving up on the binary allocation between liberty and dignity altogether. Personal development and identity, free self-realisation and autonomy have their roots in both, liberty as well as dignity. The close relation between liberty, dignity and privacy is exemplified in the constitutional construction and history of the right to privacy in some member states, which, again, has shaped the development of the right to privacy as a shared constitutional principle within the EU. The German Constitutional Court, for example, has constructed the right to privacy, which is not explicitly acknowledged in the German Fundamental Law (GFL), on the basis of the right to free development of personality—which, again, is based on the right to liberty (Art. 2(1) GFL) as well as the right to dignity (Art. 1(1) GFL) as established in the constitution.[49] In France, too, privacy protections evolved as a tool to protect the dignity of persons against excessive press coverage.[50] This underlines that liberty and dignity are two coexisting pillars, on which the right to privacy rest.

3.2 Non-Discrimination and Equal Treatment

The scope (Sect. 3.2.1), as well as the objectives and values (Sect. 3.2.2) of the rights to non-discrimination and equal treatment, are related.

3.2.1 Scope of the Rights

In EU law, the rights to non-discrimination and equal treatment are established in separate articles: Art. 21(1) CFR[51] establishes the right to non-discrimination and Art. 20 CFR[52] establishes the right to equal treatment. The ECHR, in Art. 14 ECHR

[49]BVerfG 15 December 1983, Volkszählung (1 BvR 209, 269, 362, 420, 440, 484/83); BVerfG 6 December 2005, Transsexuelle III (1 BvL 3/03).

[50]Whitman (2004).

[51]"Any discrimination based on any ground such as sex, race, color, ethnic or social origin, genetic features, language, religion or belief, political or any other opinion, membership of a national minority, property, birth, disability, age or sexual orientation shall be prohibited."

[52]"Everyone is equal before the law."

and Protocol No.12,[53] only guarantees the right to non-discrimination, Art. 20 CFR does not have an equivalent in the ECHR.[54]

The ECtHR interprets the right to non-discrimination very broadly.[55] The list of protected categories in Art. 14 ECHR and Protocol No. 12 Art. 1 is open-ended, prohibiting discriminations on grounds *"such as"* the ones explicitly named. To provide far-reaching protection, the ECtHR has interpreted the notion *"such as"* extensively and has added various categories to the ones explicitly listed.[56] According to the ECtHR, a treatment is discriminatory, if it is based on "*a personal characteristic ("status") by which persons or groups of persons are distinguishable from each other*".[57] The ECtHR emphasises that characteristics that deserve protection do not have to be *"personal in the sense that they are innate or inherent",*[58] rather, it sets a low bar according to which the characteristics only have to be "*identifiable*", or "*objective*".[59]

Despite the requirement to provide equal protection and the similar, open-ended structure of Art. 21 CFR, the European Court of Justice (ECJ) does not interpret Art. 21 CFR in the same way as the ECtHR interprets Art. 14 ECHR. The ECJ has restricted the characteristics protected by the right to non-discrimination strictly to the ones explicitly listed.[60] Rather, to provide a comparable standard of protection, the ECJ relies on the application of the right to equal treatment, Art. 20 CFR.[61] To compensate for the restrictive application of the right to non-discrimination with the application of another right, the right to equal treatment, is possible, as both rights are related. According to the ECJ, the right to non-discrimination is *"merely a*

[53]Art. 14: "The enjoyment of the rights and freedoms set forth in this Convention shall be secured without discrimination on any ground such as sex, race, color, language, religion, political or other opinion, national or social origin, association with a national minority, property, birth or other status."; Protocol 12 Art. 1: "The enjoyment of any right set forth by law shall be secured without discrimination on any ground such as sex, race, color, language, religion, political or other opinion, national or social origin, association with a national minority, property, birth or other status."

[54]Early drafts of Art. 14 still included a first paragraph establishing a right to equal treatment. On the drafting of the provision, see: Schabas (2015).

[55]For an extensive commentary on Art. 14 see: Jacobs et al. (1996).

[56]ECtHR 7 December 1976, Kjeldsen, Busk Madsen and Pedersen/Denmark (5095/71; 5920/72; 5926/72) 56.

[57]ECtHR 7 December 1976, Kjeldsen, Busk Madsen and Pedersen/Denmark (5095/71; 5920/72; 5926/72) 56; see also: ECtHR 15 March 2016, Novruk a.o./Russia (31039/11, 48511/11, 76810/12, 14618/13 and 13817/14).

[58]Para 89. ECtHR 24 May 2016, Biao/Denmark (38590/10).

[59]ECtHR 15 March 2016, Novruk a.o./Russia (31039/11, 48511/11, 76810/12, 14618/13 and 13817/14).

[60]Ward (2018); ruling that obesity does not in itself constitute a "disability" within the meaning of Directive 2000/78: ECJ 18 December 2014, FOA (C-354/13) narrow interpretation of the scope of application of Directive 2000/78/ECECJ 2 June 2016, C (Case C-122/15).

[61]Criticising the low impact of the ECJ's application: Ward (2018).

specific enunciation of the general principle of equality".[62] Both rights establish similar guarantees and the structure, in which potential breaches of the rights are examined, is identical. The ECJ explains that both *"the principles of equal treatment and non-discrimination require that comparable situations must not be treated differently and that different situations must not be treated in the same way."*[63] The assessment of potential breaches of both rights begins with the comparison of two situations, in which the ECJ establishes whether these situations are similar or different. It is only after this comparison, that the ECJ is able to establish what qualifies as a breach of the rights; whether it is treating the individuals the same way or treating them differently.

However, despite the similarities, there are also substantial differences between the right to non-discrimination and the right to equal treatment. Unlike the right to non-discrimination, the right to equal treatment does not specify what criteria must not be taken into account, when distinguishing between two situations. This makes the right to equal treatment a potentially boundless right, which applies to all differentiations in situations where fundamental rights are relevant. "Personal characteristics" or other statuses are addressed as well as any other category and type of information, as long as they relate to the *"factual and legal circumstances'*[64] of a person. To narrow down the otherwise boundless scope of the right to equal treatment, the ECJ, for two situations to be similar or different in the sense of Art. 20 CFR, requires a qualification. It applies a general test and assesses whether the factor on the basis of which two situations have been differentiated makes these situations *essentially different*.[65] In other words, the right to no equal treatment does not apply to every situation, in which a person has been treated differently despite being similar or has been treated similarly despite being different. It only applies when the difference, or correspondingly, the similarity, which has been neglected or wrongfully considered, is qualified in a way which is *"essential"*.[66] Defining what makes a difference *essential* is a complex task, which can only be accomplished with regard to the objectives and the values of the right to equal treatment.

[62]ECJ 19 October 1977, Albert Ruckdeschel & Co. and Hansa-Lagerhaus Ströh & Co. v Hauptzollamt Hamburg-St. Annen; Diamalt AG v Hauptzollamt Itzehoe (117–176 and 16–77) para 8.

[63]ECJ 11 April 2013, Soukupová (C-401/11) para 29; see also: ECJ 14 April 1994, A. v Commission (T-10/93) para 42.

[64]ECJ 14 April 1994, A. v Commission (T-10/93) para 42.

[65]ECJ 14 April 1994, A. v Commission (T-10/93) para 42.

[66]See also: ECJ 7 February 1991, Tagaras v Court of Justice of the European Communities (T-18/89 and T-24/89) para 68; the ECHR, in the application of Art. 14 ECHR requires that situations are 'relevantly' similar or different: ECtHR 29 April 2008, Burden/the United Kingdom (13378/05) para 60.

3.2.2 Objectives and Values

Considering both the similarities as well as the differences of the rights to non-discrimination and to equal treatment, a shared overarching objective can be carved out: they both have the objective to prohibit specific kinds of differentiations. More explicitly, the values which are pursued by each of the rights depend on the type of differentiations which are prohibited; on the type of categories, which are inadmissible in the differentiation of situations. Three different types of prohibitions can be distinguished. The first prohibits differentiations on the basis of categories explicitly listed within the right to non-discrimination, the second bans differentiations on the basis of categories *"such as"* the ones listed, and the third is the general prohibition established by the right to equal treatment.

As the content of the prohibition regarding differentiations on the basis of the explicitly named categories is clear, I keep the engagement with its values, for now, brief. This prohibition protects a number of groups which have been historically marginalised, are structurally disadvantaged or vulnerable. It opposes illiberal positions such as racism, sexism, antisemitism and others which have caused great harm or are, for other reasons, incompatible with the convictions underlying the CFR and the ECHR. Differentiating on the basis of the explicitly stated grounds is regarded as appalling and morally reprehensible and is therefore declared illegal. It is only the engagement with the second kind of prohibition, which makes a closer analysis of the values, which are pursued with the first prohibition, necessary.

The values pursued with the second kind of prohibition, which bans differentiations based on categories *"such as"* the ones listed, are harder to grasp. The identification of values pursued with this prohibition presupposes a closer understanding of the kind of categories, which are protected by the right to non-discrimination, even though they are not explicitly listed. Neither Art 14 ECHR nor Art 21 CFR defines, what the shared quality these categories is. The requirement which unites them is that they must be categories *"such as"* the ones listed. Hence, their shared quality can only be established in engagement with the shared nature of the categories which are listed.

In the search for the shared nature of the protected categories, the case law of the ECtHR provides little guidance. In its application of Art. 14 ECHR, the ECtHR attributes the qualification *"such as"* little importance.[67] It regards the list of protected categories as *"illustrative and not exhaustive"*[68] and extends it generously, including to categories such as employment.[69] The purpose of this approach is rather obvious: The broad interpretation allows the ECtHR to compensate for the lack of a general right to equal treatment. While, in the pursuit of encompassing protection,

[67] Schabas (2015).

[68] ECtHR 15 March 2016, Novruk a.o./Russia (31039/11, 48511/11, 76810/12, 14618/13 and 13817/14) para 90.

[69] ECtHR 27 July 2004, Sidabras a. Dziautas/Lithuania (55480/00 a. 59330/00) para 51ff.

such an interpretation is understandable, it rather obfuscates, if not intentionally blurs the line between the two rights.

In pursuit of conceptual clarity and in line with the case law of the ECJ, it is wise to attempt a more narrow definition of the shared nature of protected categories.[70] The qualifying notion "such as" derives its content from the commonalities of the explicitly named categories. The common denominator of the listed categories is that they all relate to qualities which are of great relevance to the lives of individuals and which are profound in the way that individuals cannot change (such as sex, race, colour and genetic features), or should not have to change against their choosing (such as religion, belief and political or any other opinion).[71] Hence, the assessment, whether a category is similar to the ones explicitly named, has two stages: First, it has to be established whether the category describes a quality which is of great relevance to the lives of individuals and which is profound in the way that the individual cannot influence it, or regards a fundamental choice, which an individual should not have to change. Regarding the latter type of categories, the assessment is especially complex, as it implies yet another normative question: What are individual choices, which are so profound, that they amount to be a *characteristic* of a person? Secondly, the assessment requires a normative judgement, in which it has to be established, whether a particular characteristic deserves protection in a similar way as the ones explicitly stated—and if discriminating on its basis is, therefore, also morally reprehensible. Ultimately, the prohibition to differentiate on the basis of categories "such as" the ones explicitly named pursues the same values as the original prohibition: to ban appalling and morally repressible differentiations. While the first type of characteristics protects humans from being judged on the basis of qualities which they cannot influence, the second type of characteristics protect fundamental choices in especially sensitive spheres of life.

Regarding the right to equal treatment, the identification of underlying objectives is even more complex; referring to "equality" can merely be the start of a description. "Equality" is an indeterminate concept whose content depends on the way situations are compared. If we compare two persons on the basis of categories describing their physical condition, we may come to the conclusion that they are similar. If we compare the same two persons on the basis of categories describing their clicking behaviour, movement patterns and shopping habits, we may come to the conclusion that they are different. The values pursued by the right to equal treatment can only be established after it has been defined on what basis situations are compared and differentiations are assessed. Engaging with the values of the right to equal

[70]A number of judges of the ECtHR also support a more narrow view, see the dissenting opinions in: ECtHR 27 July 2004, Sidabras a. Dziautas/Lithuania (55480/00 a. 59330/00).

[71]In her dissenting opinion, judge Thomassen is arguing in a similar way: "The principle of non-discrimination (…) refers above all to a denial of opportunities on grounds of personal choices in so far as these choices should be respected as elements of someone's personality, such as religion, political opinion, sexual orientation and gender identity, or, on the contrary, on grounds of personal features in respect of which no choice at all can be made, such as sex, race, disability and age." ECtHR 27 July 2004, Sidabras a. Dziautas/Lithuania (55480/00 a. 59330/00).

treatment, therefore, presupposes a deeper understanding of the process of comparing and differentiating.

The process of comparing as well as the decision, which kind of differences are essential, implies normative choices which cannot be deduced from the wording of the right to equal treatment. The normative choices, which are relevant here, do not so much concern the well-known conflict between formal equality and substantial equality.[72] Rather, the normative choices concern the question of what categories are considered important and what arguments are considered valid when comparing and differentiating between situations. Any comparison presupposes choices in these two regards; the selection of categories and the way of reasoning. The choices taken in these regards provide the normative foundations on the basis of which situations and individuals can be compared and on the basis of which the right to equal treatment is applied. Their importance cannot be overestimated: They predefine the results which are achieved in comparisons and which are achieved in the application of the right to equal treatment. To understand what values the ECJ's application of the right to equal treatment serves, we have to understand what the normative foundations are on the basis of which it assesses the right. The ECJ does not explain how it defines categories and how it reasons in the assessment of comparison and in its decision whether differences are essential. The maxim that "*comparable situations must not be treated differently and that different situations must not be treated in the same way*" only frames, but does not solve the problem.[73] The specification of the ECJ that "*the principle of equal treatment is breached when two categories of persons whose factual and legal circumstances disclose no essential difference*"[74] does not help much. "*Factual and legal circumstances*" are categories so broad, that they hardly narrow down relevant factors. Explanations, such as that situations have to be "*analogous*" but not "*identical*" to be considered *similar*, do not further conceptual clarity, neither.[75]

Even though not explicitly stated, the normative foundations of the assessment of the ECJ can be identified in a meta-view on its case law. They comply with what can be described as a conventional and intuitive approach. Such approach is defined by the subject matter as well as causality. The selection of categories, on the basis of which the ECJ examines situations, is shaped by the subject matter, and the way it attributes relevancy to these categories is defined by causality. If the subject matter is road safety, the physical preconditions of a driver, especially his or her eyesight, are important categories.[76] Why physical preconditions are important when judging a

[72]See for example: Barnard and Hepple (2000).

[73]ECJ 11 April 2013, Soukupová (C-401/11) para 29; see also: ECJ 14 April 1994, A. v Commission (T-10/93) para 42.

[74]ECJ 14 April 1994, A. v Commission (T-10/93) para 42.

[75]Schabas (2015).

[76]ECJ 22 May 2014, Glatzel (C-356/12).

person's capacity to drive, can be explained on the basis of causality: If a driver does not have good vision, he or she cannot manoeuvre a vehicle safely.[77]

With regard to the original question of what values the right to equal treatment pursues, it can be summarised that the right to equal treatment pursues the appropriate differentiation between individuals. What differentiations are regarded as appropriate is determined by underlying normative choices. In the approach of the ECJ, differentiations are appropriate, if they respond to the subject matter and are established on the basis of causality.

4 Allocating and Responding to the Challenges

In this final part of the paper, I assess the implication of behavioural profiling on each of the rights, taking into account their scope as well as their objectives and the pursued values. I allocate the challenges before suggesting ways to respond to them. I first asses the right to privacy, (Sect. 4.1) before turning to the right to non-discrimination and the right to equal treatment (Sect. 4.2).

4.1 Group Privacy as One Guarantee of the Right to Privacy

It has been pointed out that the ECtHR and the ECJ do not establish a conclusive definition of the right to privacy. Rather, they establish various guarantees which are regarded necessary to protect privacy. In face of novel technologies, the open-ended nature of the right to privacy is a merit rather than a deficiency. It allows to adapt the scope of the right in a way which responds to new challenges, effectively realising its objectives and protecting the values which are pursued with it.

The unique potential of the right to privacy to address novel concerns for privacy can be illustrated when contrasted with the right to the protection of personal data (Article 8 CFR[78]). The right to data protection is conclusively defined: it protects data which is "personal," in the sense that it relates to an "identified' or "identifiable' natural person.[79] While this definition is unambiguous and clear, it is also rigid and incapable to respond to new threats which emerge from new forms of data analysis.[80] Scholars have demonstrated that data can violate the privacy of individuals, even if it isn't personal.[81] The more advanced machine learning data analysis is, the less it

[77]The casual reasoning is also apparent in the justification: ECJ 22 May 2014, Glatzel (C-356/12) para 54.

[78]Article 8(1): "Everyone has the right to the protection of personal data concerning him or her."

[79]See the similar definition in Directive 2016/680 of the EU (27 April 2016), para 21.

[80]Arguing the opposite: Lynskey (2014).

[81]See the explanations in Sect. 2.2.

relies on data relating to "identified' or "identifiable' persons. Therefore, it is apprehended that the right to data protection will not provide sufficient protection in the future. The scope of the right to privacy, conversely, with its open-ended structure, is capable of adapting to technological change.

The right to privacy finds its foundation in human liberty as well as dignity. In its case law, the ECtHR has made clear, that it protects personal development, free self-realisation and personal autonomy. The question scholars have to answer when wondering whether the right to privacy applies to machine learning based data analysis, is not so much, whether one of the established guarantees applies and whether, therefore, machine learning based data analysis falls into the scope of the right to privacy. The question is, contrarily, whether one of the objectives and values of the right to privacy is affected, and therefore, a guarantee should apply. The scope of the right can and must be redefined in light of its objectives and the values it pursues. It has been convincingly argued that machine learning based data analysis affects personal development, free self-realisation and personal autonomy, and with it individual liberty and dignity, even if it is not based on personal information.[82] The currently recognised guarantees of the right to privacy have to be complemented by a new guarantee which effectively responds to the novel threats.

The challenge privacy scholars face is to identify what effects the machine learning based generation of information has, how these effects restrict something that the right to privacy is supposed to protect and, if so, what guarantees are necessary in order to prevent these effects. The efforts of scholars, engaging with the notion of Group Privacy, are promising. However, conceptual clarity is crucial here, too. In this debate, the distinction between the scope of the right and its objectives and values is essential. The notion of "Group Privacy" is used on both sides of this distinction, in reference to the values of the right as well as to its scope. Some scholars argue that "Group Privacy' is a value which requires protection.[83] The identity of groups, it is argued, is similarly constructed to the identity of individuals and therefore is a value worth protecting.[84] This approach, evolving the values and objectives of the right to privacy rather than its scope, threatens to dilute the foundations of the right to privacy. The right to privacy has its foundations in the liberty and dignity of the individual, and it protects personal development, self-determination and identity. In the versatile history of the right, these values have been the stabilising factors on which the dynamic evolution of the scope of the right has rested.

The approach of other scholars, who regard Group Privacy as one guarantee, which should complement the scope of the right, is more promising. Scholars argue that the protection of information about groups is necessary in order to protect the individual effectively.[85] As it is often group affiliations, which determine the

[82]Idem.

[83]Floridi (2017).

[84]Floridi (2011, 2017).

[85]Kammourieh et al. (2017).

perception and treatment of individuals in behavioural profiling, for the sake of the individual, groups must be protected, too. This approach stands in line with the dynamic interpretation of the right to privacy, whose guarantees evolve, while the objectives and values remain the same.

4.2 Not Non-Discrimination But Equal Treatment

The innumerable and fluent groups formed on the basis of, at least seemingly, trivial behaviour mostly do not match with the explicitly listed categories of the right to non-discrimination. The only way to apply the right to such groups would be to interpret the notion *"such as"* very broadly and to append a larger number of categories to its open-ended list, even though these categories would be only remotely, if at all related to the categories which are explicitly listed. The case law of the ECtHR goes in this direction, protecting all *identifiable, objective or personal characteristics.*[86] Taking into account the objective and the values of the right to non-discrimination, this is not the right way forward.

The right to non-discrimination is for any pluralistic society an indispensable right, whose purpose is not to prohibit all kinds of, but only the most appalling and morally reprehensible differentiations. It grants a unique status to individuals who belong to historically marginalised, structurally disadvantaged or vulnerable groups which require extraordinary protection. It protects individuals from being judged on the basis of qualities which they cannot influence or on the basis of defining choices in sensitive spheres of life. Extending the scope of the right to non-discrimination through a broad interpretation of the notion *"such as"* would water down its distinguished legal and moral force. Discriminations which happened on the basis of random categories would be granted the same status as morally reprehensible discriminations. An extensive interpretation would further the protection of all individuals in all spheres of life at the cost of the distinguished protection of vulnerable groups and on the cost of the unique protection of sensitive spheres of life and is, therefore, to be rejected.

Instead, the notion *"such as"* should be interpreted narrowly, and include only categories which are similar to the ones explicitly named. Categories are only similar, when, firstly, they relate to qualities which are of great relevance to individuals' lives and which are profound in the way that individuals cannot change or should not have to change them against their choosing. Regarding the latter kind of categories, it has to be established that choices taken within these categories have such a profound effect on the life of an individual that they should be regarded as characteristics. Secondly, categories are only similar to the ones listed when differentiating on their basis is also morally reprehensible. Interpreted this way, the right

[86]ECtHR 15 March 2016, Novruk a.o./Russia (31039/11, 48511/11, 76810/12, 14618/13 and 1817/14).

to non-discrimination does not apply to most categories used and groups formed in behavioural profiling. The categories and groups which are taken as a basis to differentiate between individuals in behavioural profiling are mostly too granular, too random and too fluid to be considered similar to the explicitly protected categories and groups of the right to non-discrimination.[87] Categories such as clicking behaviour, mouse moving patterns or the speed with which one scrolls through one's social media feed do not relate to qualities which are of great relevance to individuals' lives and which are profound in the way that individuals cannot, or should not have to change them against their choosing. These categories are not similar to the characteristics explicitly protected, in fact, they are not characteristics at all. Neither is differentiating on their basis per se morally reprehensible. They are not categories such as the ones explicitly named, and the right to non-discrimination does not and should not apply to them. Indeed, a narrow interpretation limits the scope of the right to non-discrimination, despite the fact that behavioural profiling poses new threats and makes further protection necessary. However, the sorrow that the right to non-discrimination fails to provide protection against behavioural profiling is mitigated by the insight, that it never had the purpose to address differentiations which happen on the basis of random categories in the first place—and that a restrictive interpretation allows it to provide protection where it is most needed.[88]

The right to equal treatment, conversely, carries great, even though little acknowledged potential to provide an adequate normative framework for behavioural profiling. The scope of the right to equal treatment flawlessly applies to behavioural profiling, as the right to equal treatment, in contrast to the right to non-discrimination, addresses all kinds of differentiations. Its scope is not, a priori, limited to a certain set of categories, to vulnerable groups, to characteristics or to sensitive choices which shape the lives of individuals. Neither does it only address differentiations which are morally reprehensible. The right to equal treatment applies to differentiations, which have been made on any kind of category. It protects any kind of group, not only those which are formed on the basis of protected characteristics, such as gender, race or ethnicity, but also groups which are clustered together on the basis of random and trivial activities.

It is no coincidence that the scope of the right to equal treatment applies so impeccably to behavioural profiling. It is the very objective of the right to equal treatment to assess practices such as behavioural profiling. Both behavioural profiling and the right to equal treatment are, in their essence, about one and the same matter: differentiating between individuals. Behavioural profiling can be regarded as a technology whose purpose it is to differentiate between individuals: First it is analysed who is a careful driver and who is not, who will dispose over sufficient financial resources and who will not, who will be in good health and who will be ill. Then the treatment which individuals receive is determined on the basis of this analysis. Every individual receives a personalised treatment, the differences in the

[87]For the type of protected groups, see: Kammourieh et al. (2017).

[88]Leese (2014).

treatments mirroring the differences detected in the analysis: Who should receive an offer for an insurance or for a loan, and under what conditions, and who should not receive an offer at all? The right to equal treatment, correspondingly, has the purpose of assessing differentiations. The questions which are answered on the basis of the right to equal treatment are: Is it legal to charge one driver higher insurance rates than another driver, to deny a loan to one person but to grant it to another one, to give someone access to health insurance while denying it to someone else? Put more generally, the application of the right to equal treatment consists in the question: Do differences between individuals justify the differences in treatment which they receive? Scholars have convincingly described machine learning based data analysis as a new paradigm of generating knowledge.[89] Machine learning based behavioural profiling is, more precisely, a new paradigm of differentiating between individuals. It is the right to equal treatment, which provides the normative framework in order to respond to this new paradigm of differentiating.

However, the application of the right poses challenges. The first challenge concerns the scope of the right and is to decide whether individuals are *essentially different* or *equal* in the sense of the right. According to the ECJ, the right to equal treatment only applies, when two situations, which are essentially different, are treated the same, or when situations, which are essentially the same, are treated differently. The hard question scholars and courts have to answer in the assessment of the scope of the right is: If individuals are clustered into different groups by machine learning algorithms, does that make them *essentially different*—meaning that the right to equal treatment does not apply, when they are treated *differently*? Put differently, if the behaviour of two individuals correlates with different outcomes, does that mean that they do not have the right to be treated the same? If we do regard individuals as similar, despite the fact that their data correlates with different outcomes, the scope of the right to equal treatment applies—and a second hard question, concerning the justification of the discrimination, arises. The hard question concerning the justification is: Does information generated by machine learning algorithms provide a due reason to evaluate and treat some individuals worse than others? Put differently: Do correlations between seemingly random categories justify disadvantages?

The questions concerning the scope as well as the justification of the right to equal treatment originate in one and the same problem and can be discussed in one and the same normative debate. The shared problem concerns implicit choices, which underlie comparisons and differentiations. The overarching debate, on the basis of which the questions relating to the scope and the justification of the right can be answered, must make these implicit choices obvious, must formulate the challenges which behavioural profiling brings with it when making these choices, and must define parameters on the basis of which these choices can be made in the future.

As it has been argued earlier that, to realise the existence and understand the significance of the implicit choices, one has to scrutinise the process of comparing.

[89]Hildebrandt and Koops (2010).

Comparisons rest on normative foundations, they presuppose choices regarding the selection of categories and the way of reasoning. The normative foundations predefine the results which are going to be achieved in comparisons. As the right to equal treatment consists, in its essence, in an assessment of comparisons and differentiations, its application presupposes and is defined by these implicit choices, too. Behavioural profiling poses a great challenge to the application of the right to equal treatment, as the normative foundations, on the basis of which it operates, and hence compares and differentiates between individuals, are fundamentally different from the normative foundations, on the basis of which humans intuitively and courts conventionally compare and differentiate. The machine learning based approach and the conventional human approach differ in both of the implicit choices, the selection of categories as well as the logic according to which the relevancy of the categories is assessed. The intuitive and conventional approach is defined by the subject matter as well as causality. When humans assess if a person is a safe driver, they take into account factors which they believe are causally linked to the characteristic of being a safe driver, such as a person's driving history or his or her physical preconditions. In the way machine learning algorithms compare and differentiate, conversely, neither the subject matter, nor causality are relevant. Rather, machine learning algorithms consider all available categories and they attribute relevancy to categories on the basis of patterns and correlations. When machine learning algorithms assess if a person is a safe driver, they take into account all kinds of data points, such as clicking behaviour, mouse moving patterns or the speed with which a person scrolls through his or her social media account. It is statistical rather than causal relations, that make valid arguments in behavioural profiling. The normative foundations, on the basis of which humans and courts compare and differentiate, and on the basis of which the right to equal treatment is conventionally applied, are antithetical to the way machine learning algorithms do.

The great challenge that fundamental rights scholars and courts are confronted with in the application of the right to equal treatment is to develop standards which can guide the decision whether and when evaluations and predictions generated with machine learning based algorithms constitute legitimate reasons to differentiate between individuals. They have to engage in the question if and how to accommodate the antithetical normative foundations on the basis of which machine learning algorithms compare and differentiate within the conventional assessment of the right to equal treatment. Fundamental rights scholars have to develop a normative framework to judge whether differentiations on the basis of correlative relations between seemingly random categories can be assessed.

The difficulties occurring in the application of the right to equal treatment are manifestations of a greater challenge, which transcends the application of fundamental rights. The conventional way of courts to assess differentiations is deeply rooted in societal understandings of how to make sense of the world. The normative foundations of the application of the right to equal treatment—the selection of categories and the way of reasoning—are not only the basis for comparisons and differentiations. The choices in these two respects, more generally, define the way knowledge is generated, the logic on the basis of which the world is perceived and explained. Machine learning algorithms do not only challenge the way we compare and differentiate, but, more generally, the way we describe and make sense of the world.[90] The examination of the subject matter on the basis of causality versus the analysis of all available information on the basis of correlation is not only the battle line within the application of the right to equal treatment—it is the fundamental conflict with which Big Data confronts society. The problems within the application of the right to equal treatment are an expression of a larger epistemological challenge. The larger epistemological challenge consists in the development of a response to machine learning algorithms new paradigm of generating knowledge.

The current way in which the right to equal treatment is applied can be regarded as an expression of a societal consensus of how to make sense of the world. According to this consensus situations should be described and judged on the basis of causal relations within the subject matter. It is because this consensus is challenged, that the normative foundations of the application of the right to equal treatment tremble. Fundamental rights scholars cannot conclusively solve the challenges within the application of the right to equal treatment as they cannot, through legal interpretation, solve the overarching epistemological challenge. The right to equal treatment does not define the logic, on the basis of which comparisons and differentiations should be assessed but presupposes it. The task to develop solutions for the application of the right to equal treatment to behavioural profiling can not be solved through legal reasoning and within the legal debate alone. It is not only fundamental rights lawyers, but society at large, that has to decide if and how to alter its conventional way to explain the world and to differentiate, in order to accommodate the new paradigm of generating knowledge, which Big Data brings with it. Solving the overarching challenge requires a societal debate on the question of how to adapt our conventional and intuitive way of perceiving and explaining the world, our conventional logic, to the new way in which machine learning algorithms generate information. And it is only on the basis of this larger debate, that fundamental rights scholars and courts can decide if and how to alter the normative foundations of the right to equal treatment, if and how to alter the conventional way in which they compare and assess differentiations. Fundamental rights scholars can and must contribute to this larger societal debate, but they cannot predefine its results.

5 Conclusion

It is important to recall why fundamental rights matter in the debate on behavioural profiling, even though in western democracies, it is mostly companies who use it. Fundamental rights matter, first, because they do not only entail negative

[90] Hildebrandt and Koops (2010) and Mayer-Schönberger and Cukier (2013).

obligations, the freedom of state interference, but also positive obligations. States in general and legislators in particular have the duty to actively promote the protection of fundamental rights, even in the relation among privates.[91] In order to concretise the positive obligations, it is crucial to know, what and how fundamental rights are affected by behavioural profiling. Secondly, fundamental rights perspectives can inform societal, value based debates. Fundamental rights assessments and societal, value based debates are intertwined: The assessment of fundamental rights is influenced by societal understandings and ideals, while, at the same time, the fundamental rights perspective shapes societal understandings and ideals. Fundamental rights at the same time mirror and coin societal values. Conceptual clarity in the assessment of fundamental rights, therefore, furthers conceptual clarity within the societal, value based debate.

Also, we should be well aware of what is at stake when determining on the application of the right to equal treatment to behavioural profiling. The consequences of the way in which we apply the right to equal treatment to behavioural profiling are potentially drastic. If courts maintain their conventional way to apply of the right to equal treatment, they are likely to declare large parts of behavioural profiling illegal. If machine learning algorithms treat two individuals differently on the basis of correlative relations between seemingly random categories, the scope of the right to equal treatment would apply; and correlative relations would not be regarded as suitable reasons to justify disadvantages. If a public insurer adapted its prices on the basis of machine learning based analysis of clicking behaviour, typing patterns and scrolling speed, the right to equal treatment would apply and a justification could often not be obtained. The same would be true for any situation, in which random categories and correlative relations were used in the data analysis, which is the case in practically all of machine learning based behavioural profiling. Such an application of the right to equal treatment, however, could block the innovative potential of machine learning algorithms. Besides its low costs and high efficiency, their innovative potential consists in the fact, that they expose relations which cannot be identified in conventional assessments of the subject matter on the basis of causality.[92] Maintaining their conventional standards, courts and legislators, complying with their duty to protect fundamental rights through legislation governing the relation among privates, could ultimately limit the use of behavioural profiling to situations in which the generated results are explainable.[93] Whether this is desirable or not, is a challenging question. A persuading answer should reconcile the conventional standards of justice with the innovative potential of machine learning algorithms.

Further-on, much is won when labelling the assessment whether two situations are similar or different an assessment of legitimacy rather than relevancy. The more

[91]For the right to privacy, see: ECtHR 13 June 1979, Marckx/Belgium (6833/74).

[92]Mayer-Schönberger and Cukier (2013).

[93]For a suggestion how to incorporate the explainability into the technology, see: Wachter et al. (2018).

common and intuitive label of relevancy purports objectivity, even though it is not given. The label of relevancy obfuscates the normative choices on which the comparisons, which define the scope of the right to equal treatment, rest. The logic underlying the conventional application of the right to equal treatment—identifying causal relations of factors within the subject matter—is only one way to perceive and make sense the world. The underlying consensus is indeed so broad, that implicit choices became almost invisible; appearing as self-evident truths rather than normative choices. However, when confronted with the new logic of machine learning data analysis, its particularities become apparent and a fundamental question, which precedes the matter of relevancy, arises. The question is not whether a category is relevant for a differentiation, but what standards are applied to decide whether a category is relevant or not. The label of legitimacy makes the normative choices, which the answer to this question implies, obvious—and open for discussion. Legitimacy is a multifaceted term, which can encompass both, the constitutional as well as the societal debate on an adequate normative framework for machine learning based differentiations.

The important even though not solely decisive fundamental rights debate of the matter can begin with a normative statement, which finds its basis in human dignity, liberty as well as equality. Its content only becomes fully obvious in consideration of the factors which it excludes: people should be evaluated and treated on the basis of their actions. Not on the basis of their religion, sex or race (the right to non-discrimination). Not on the basis of characteristics which are of great relevance for their lives and are profound in a way that they cannot or should not have to change them (categories "such as" the ones explicitly named in the right to non-discrimination). The identification of such kind of negative amendments can further the debate. Such amendments can be nuanced, such as: not on the basis of his or her movement patterns? Not on the basis of people who surround him or her? Or, they might take a more general and radical shape: not on the basis of statistical correlations of their behaviour? Not on the basis of machine learning based analysis which expose information the individual did not consent to share?

Also, the debate has to address questions of what trade-offs a society is willing to make in order to benefit from the innovative potential of machine learning based data analysis.[94] Trade-offs concern the protection of privacy, the accuracy of data analysis and with it the just treatment of individuals. While data analysis on the basis of big data sets and statistical patterns might increase the accuracy of data analysis at large, sometimes in this analysis, individuals fall through the grid.[95] Imperfect data sets lead to imperfect results. Even if an encompassing evaluation of an individual's behaviour, be it on the basis of correlation or causality, promises great results, an evaluation on the basis of limited information might lead to poor results. The less data is available about individuals, the more likely will machine learning algorithms generate incorrect information about them. If the normative

[94]On the value of big data analytics see: Mayer-Schönberger and Cukier (2013).

[95]For an understanding of how that happens, see: Hildebrandt and Koops (2010).

foundations of the right to equal treatment are adapted in a way that the right no longer prohibits differentiations on the basis of correlative relations between random categories, thus enabling behavioural profiling, this creates an incentive for individuals, to avoid disadvantageous treatments on the basis of incorrect information, to share more data. Individuals would have to renounce on privacy in order to be treated justly. Whether prioritising accuracy and innovative potential over privacy is path worth pursuing is a difficult question, which can be debated controversially. What is crucial to acknowledge however, is that these overarching choices are defining for the application of the right to equal treatment, too.

These are some of the foundations form which a debate on behavioural profiling can depart and some of the questions, to which fundamental rights scholars and society at large have to develop answers. The choices taken in this debate will determine the way in which the right to equal treatment will be applied in the future—and on the place that machine learning based profiling will take in society.

References

Barnard C, Hepple B (2000) Substantive equality. Cambridge Law J 59:562

Barocas S, Selbst AD (2016) Big data's disparate impact. California Law Rev 104:671

Corbett-Davies S et al (2016) A computer program used for bail and sentencing decisions was labelled biased against blacks. It's actually not that clear. The Washington Post, 17th October 2016. https://www.washingtonpost.com/news/monkey-cage/wp/2016/10/17/can-an-algorithm-be-racist-our-analysis-is-more-cautious-than-propublicas/?noredirect=on&utm_term=.9c25816db76c. Accessed Apr 2019

Floridi L (2011) The informational nature of personal identity. Minds Mach 21:549

Floridi L (2014) Open data, data protection, and group privacy. Philos Technol 27:1

Floridi L (2017) Group privacy: a defense and an interpretation. In: Taylor L et al (eds) Group privacy. Springer, Dordrecht

Gonçalves ME (2017) The EU data protection reform and the challenges of big data: remaining uncertainties and ways forward. Inf Commun Technol Law 26:90

Hacker P (2018) Teaching fairness to artificial intelligence: existing and novel strategies against algorithmic discrimination under EU law. Common Mark Law Rev 55:1143

Hacker P, Wiedemann E (2017) A continuous framework for fairness. arXiv.org. http://arxiv.org/abs/1712.07924

Harris D et al (2009) Law of the European Convention on human rights. Oxford University Press, New York

Hildebrandt M (2006) Profiling: from data to knowledge. Datenschutz und Datensicherheit 30:548

Hildebrandt M, Gutwirth S (eds) (2008) Profiling the European citizen. Springer, Dordrecht

Hildebrandt M, Koops EJ (2010) The challenges of ambient law and legal protection in the profiling era. Mod Law Rev 73:428

Jacobs FG et al (1996) The European Convention on human rights. Clarendon Press, Oxford

Kammourieh L et al (2017) Group privacy in the age of big data. In: Taylor L et al (eds) Group privacy. Springer, Dordrecht

Kim P (2017) Data-driven discrimination at work. William Mary Law Rev 58:857

Kokott J, Sobotta C (2015) Protection of fundamental rights in the European Union: on the relationship between EU fundamental rights, the European Convention and national standards of protection. Yearb Eur Law 34:60

Leese M (2014) The new profiling: algorithms, black boxes, and the failure of anti-discriminatory safeguards in the European Union. Secur Dialogue 45:494
Lynskey O (2014) Deconstructing data protection: the "added-value" of a right to data protection in the EU legal-order. Int Comput Law Q 63:569
Mantelero A (2016) Personal data for decisional purposes in the age of analytics: from an individual to a collective dimension of data protection. Comput Law Secur Rev 32:238
Mayer-Schönberger V, Cukier K (2013) Big data: a revolution that will transform how we live, work and think. John Murray, London
Mittelstadt B (2017) From individual to group privacy in big data analytics. Philos Technol 30:475
Mittelstadt B et al (2016) The ethics of algorithms: mapping the debate. Big Data Soc 3:1
Noble SU (2018) Algorithms of oppression: how search engines reinforce racism. New York University Press, New York
Roberts JL (2015) Protecting privacy to prevent discrimination. William Mary Law Rev 56:2132
Rubinstein IS (2013) Big data: the end of privacy or a new beginning? Int Data Priv Law 3:74
Schabas WA (2015) The European convention on human rights: a commentary. Oxford University Press, Oxford
Schermer BW (2011) The limits of privacy in automated profiling and data mining. Comput Law Secur Rev 27:45
Solove DJ (2002) Conceptualizing privacy. California Law Rev 90:1087
Taylor L et al (2017a) Introduction: a new perspective on privacy. In: Taylor L et al (eds) Group privacy. Springer, Dordrecht
Taylor L et al (2017b) Group privacy. Springer, Dordrecht
Van der Sloot B (2017) Do groups have a right to protect their group interest in privacy and should they? Peeling the onion of rights and interests protected under article 8 ECHR. In: Taylor L et al (eds) Group privacy. Springer, Dordrecht
Wachter S, Mittelstadt B (2018) A right to reasonable inferences: re-thinking data protection in the age of big data and AI. Columbia Bus Law Rev. https://papers.ssrn.com/sol3/papers.cfm?abstract_id=3248829
Wachter S et al (2017a) Transparent, explainable, and accountable AI for robotics. Sci Robot 2:1
Wachter S et al (2017b) Why a right to explanation of automated decision-making does not exist in the general data protection regulation. Int Data Priv Law 7:76
Wachter S et al (2018) Counterfactual explanations without opening the black box: automated decisions and the GDPR. Harv J Law Technol 31:841
Ward A (2018) The impact of the EU Charter of fundamental rights on anti-discrimination law: more a whimper than a bang? Cambridge Yearb Eur Legal Stud 20:32
Wei Y et al (2015) Credit scoring with social network data. Mark Sci 35:234
Whitman JQ (2004) The two western cultures of privacy: dignity versus liberty. Yale Law J 113:1151
Zarsky T (2016) The trouble with algorithmic decisions: an analytic road map to examine efficiency and fairness in automated and opaque decision making. Sci Technol Hum Values 41:118
Zuboff S (2019) The age of surveillance capitalism: the fight for a human future at the new frontier of power. Public Affairs, New York

The Black Box on Trial: The Impact of Algorithmic Opacity on Fair Trial Rights in Criminal Proceedings

Francesca Palmiotto

Contents

Abstract Algorithms are increasingly used in criminal proceedings for evidentiary purposes (*e.g.* GPS position, for DNA analysis or as means of obtaining digital evidence) and for supporting decision-making (*e.g.* software for assessing the risk of recidivism of individuals). In a worrying trend, these tools are still concealed in secrecy and opacity preventing the possibility to understand how their specific output has been generated. This chapter focuses on the legal challenges triggered by algorithmic opacity in criminal proceedings. The chapter argues that algorithms may contain miscodes, opacity in algorithms keeps such miscodes hidden, and, as a result, algorithmic opacity impacts fair trial rights.

F. Palmiotto (✉)
European University Institute, Fiesole, Italy

M. Ebers, M. Cantero Gamito (eds.), *Algorithmic Governance and Governance of Algorithms*, Data Science, Machine Intelligence, and Law 1,
https://doi.org/10.1007/978-3-030-50559-2_3

1 Introduction

In 2013 in Wisconsin, Eric Loomis was charged with five criminal counts related to a drive-by shooting.[1] The trial court sentenced the defendant to six years of imprisonment and five years of extended supervision and referred to the COMPAS (Correctional Offender Management Profiling for Alternative Sanctions) assessment.[2] COMPAS is an automated risk-assessment software used to assess a criminal defendant's likelihood of becoming a recidivist. Three years later, an independent analysis of the software found that "black defendants were far more likely than white defendants to be incorrectly judged of possessing a higher risk of recidivism, while white defendants were more likely than black defendants to be incorrectly assessed as low risk".[3] The software was thus perpetrating a form of indirect discrimination.

In 2015, Volkswagen was embroiled in the 'emission-cheating scandal'.[4] The company admitted to having installed illegal defeat devices in its diesel vehicles. The software was designed to make the vehicles meet the pollution standards when monitored in tests, whereas during actual use their emissions exceeded by up to 40 times the legal limits in the US.

In 2019, an Italian newspaper reported that: "more than a thousand Italians have been mistakenly intercepted by a State-sponsored hacker".[5] Researchers from Security Without Borders and Motherboard[6] studied the malware in question, known as Exodus, which had been sold to the Italian government by a private company for carrying out criminal investigations, and found that the software was missing a procedure of validating its targets due to an error in its program code. Exodus did not properly check whether the phone in which it was installed was intended to be legally intercepted or not. Since Italian law enforcement is legally prohibited from hacking personal devices without a court order, many innocent citizens had been spied on by the State "by mistake".[7]

What do these examples have in common? They all involve an algorithm, containing hidden flaws, that impaired individual rights. The situation in which "we can observe inputs and outputs of an algorithm, but we cannot tell how one becomes the other",[8] is associated with the metaphor of a 'Black Box' or with the

---

[1] *State v Loomis, No 2015AP157-CR (Wis Ct App Sept 17, 2015), 2015 WL 1724741* (Supreme Court of Wisconsin).

[2] Freeman (2016), p. 75.

[3] Angwin and Larson (2016).

[4] Hotten (2015).

[5] Longo (2019).

[6] Maiberg et al. (2019) and Security without Borders (2019).

[7] According to publicly available statistics, as well as confirmation from Google, most of the apps containing the malware have been installed few dozen times, with one case reaching over 350 downloads. See for more Security without Borders (2019).

[8] Pasquale (2015).

notion of 'algorithmic opacity'.[9] This chapter aims to analyse the legal challenges triggered by the use of algorithms in criminal proceedings, highlighting the inverse relationship between algorithmic opacity and contestability.

This work builds on three hypotheses: algorithms may contain miscodes[10] (H^1), opacity keeps such miscodes hidden (H^2) and, as a result, algorithmic opacity impacts fair trial rights (H^3). After a theoretical explanation of these three hypotheses, a concrete case-study is presented: the Harm Assessment Risk Tool (hereafter HART), the first intelligent decision-support algorithm used by UK police forces for assessing the risk of future offending by arrestees. Conclusively, this chapter will argue that the opacity surrounding these software represents an obstacle and a threat to fair trial rights and principles, as it prevents the possibility to contest algorithmic models and decisions.

2 Algorithms May Contain Miscodes: The Use of Algorithms in Criminal Proceedings

Algorithms are increasingly used at every stage of criminal proceedings. At the pre-trial stage, for instance, software can assist police in investigation as means of obtaining evidence (*e.g.* malware, spyware and viruses that can be installed for controlling and having access to the camera, the memory unit and even to the microphone of the intercepted device. Another example is facial recognition software that can be helpful for identifying suspects of crimes). At the trial stage, algorithms are used for evidentiary purposes (*e.g.* the GPS position of a suspect, emails or online-chat) and even for sentencing (*e.g.* risk-assessment algorithms).

Algorithms are a longstanding concept that can be defined as "any well-defined computational procedure that takes some value, or set of values, as input and produces some value, or set of values, as output. An algorithm is thus a sequence of computational steps that transform the input into the output".[11] In the broadest sense, "an algorithm is simply a set of instructions. Cooking recipes, instructions on how to play a game, and walking directions are all everyday illustrations of what could be called algorithms".[12] Based on the instructions contained in the code, an input can be transformed into different sets of outputs.

[9]For the purpose of this chapter, the words "Black Box", "Opaque Algorithms" and "Algorithmic Opacity" will be used as synonyms.

[10]In this chapter, I will use 'miscode' as an umbrella term that encompasses errors, bias or manipulation in algorithms that impact the resulting output in terms of accuracy, validity and legality.

[11]Cormen et al. (2009), p. 5. For a definition of algorithms as "encoded procedures for transforming input data into a desired output, based on specified calculations" see Gillespie (2014).

[12]'Written evidence submitted by Google (ADM0016)' in House of Commons Science and Technology Committee (2018).

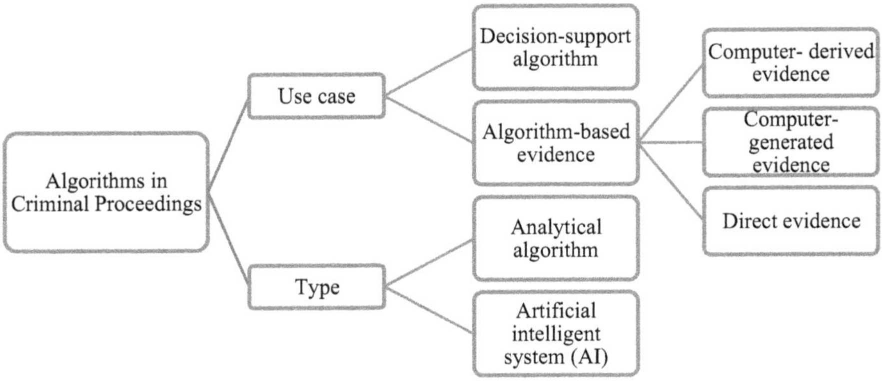

Fig. 1 Classification of algorithms in criminal proceedings

Due to the varied use of algorithms, this chapter proposes a twofold grouping (see Fig. 1 *infra*) that should be borne in mind: (1) decision-support algorithms and (2) algorithm-based evidence. The first type is used to inform human decision-making at the pre-trial, trial and post-trial stages (*e.g.* crime data mining, risk-assessment software for bail hearings, sentencing, probation and parole). The second type is used for evidentiary purposes and includes computer-derived evidence (*e.g.* GPS positions) and computer-generated evidence (*e.g.* breathalysers, DNA analysis or computer simulations of physical events[13]). Furthermore, in some cases, the algorithm itself can constitute direct evidence, such as in the famous example of the "Dieselgate scandal" in which the algorithm of the defeat device was directly relevant to demonstrating the alleged illegal behaviour of the car company.

A second relevant distinction can be made between simpler analytical algorithms and artificial intelligence systems (hereafter AI). The first involves "relationships between inputs and outputs crafted by hand. These are often rule-based systems, such as flowcharts, where the steps, methodologies and outcomes can be traced to pre-programmed instructions inputted by a human".[14] The second are "more advanced algorithmic systems which include machine learning approaches, are used with problems where pre-existing rules or theories do not capture the desired input-output relationships well. [...] These systems are often referred to as non-parametric, as the parameters that define the input-output transformation are not defined in advance, but derived from the data or complex computational simulations".[15] In criminal proceedings, AI has become more and more effective in

[13]Jenkins (1975), p. 600.

[14]The Law Society of England and Wales (2019).

[15]The Law Society of England and Wales (2019).

combating dynamically evolving cyber-attacks,[16] in providing relevant sources of evidence,[17] and even in supporting the decisions of judges or police officers.[18]

In algorithmic systems, there is a direct correlation between the input, the algorithm and the output. However, this interplay can be affected by errors, bias or manipulation that alter the results. In this chapter I will use 'miscode', that is the coding of a program wrongly or mistakenly, as an umbrella term that refers to any flaw that impacts the output in terms of accuracy, validity and legality. The presence of miscodes in algorithmic systems will demonstrate that "the reliability of computer programs should be proven rather than presumed".[19]

Miscodes can enter the algorithm in different ways and can present themselves at any stage of the lifecycle. They can originate from the input[20] or they can be contained in the code of the program.[21] Moreover, a miscode can also derive from human intervention,[22] for instance from the performance of the process by a person.[23]

In the light of a preliminary analysis of relevant case studies, I propose a threefold categorization of miscodes: technical, scientific and legal.

Technical Miscode A technical miscode refers to algorithms containing errors that affect the *accuracy*[24] of the output. In the majority of cases, a technical miscode is unintentionally made by the programmer of the software. It can also result from a 'software rot', that is "a deteriorating and outdating of code over time".[25] Moreover, an intelligent software "might also teach itself to lie as a strategy for achieving a goal".[26] One could consider the following examples:

---

[16]Dilek et al. (2015), pp. 21–22.

[17]An interesting example is the use of conversational agents, such as Siri, Alexa or Google Home, as source of evidence since they can gather data about the daily routine and activities of individuals. See Feldman (2018), Sauer (2017) and Vowles and Story Carter (2018).

[18]Generally they are used for risk assessments and are based on machine learning techniques that aim to identifying common patterns and rules from huge amounts of data. See infra §4 as an example. For more information see Green (2018).

[19]Chessman (2017).

[20]The input data fed into the machines may be inaccurate or biased. Moreover, in the specific case of machine learning (a particular type of algorithm that builds a mathematical model from sample data) the use of inappropriate or insufficient training data can have a huge impact on the structure of the algorithms and, consequently, on the output.

[21]Oxford and Nottingham Universities warned that, as the complexity of algorithmic applications increases: "So do the inherent risks of bias, as there is a greater number of stages in the process where errors can occur and accumulate. These problems are in turn exacerbated by the absence of oversight and effective regulation".

[22]See in this sense the concept of algorithmic system provided by Ananny and Crawford (2018): an algorithm is "not just code and data but an assemblage of human and non-human actors".

[23]Horsman (2019).

[24]The state or quality of being correct or precise.

[25]Roth (2016).

[26]Roth (2016).

Timestamps and EnCase [Algorithm-Based Evidence]
EnCase is a popular forensic tool used to collect computer-derived evidence, including timestamps. A timestamp is an information about the time of creation, access and modification of a certain file. Although the reliability of EnCase has rarely been questioned, relevant research has shown that by using an undocumented Windows system call, instead of another system call, all timestamps could be undetectably modified.[27]

STRmix DNA [Algorithm-Based Evidence]
STRmix DNA is an automated forensic software that uses complex mathematical formulae to examine the statistical likelihood that a certain genotype can be attributed to one specific individual over another. This instrument is used in several jurisdictions in the US, Canada and Europe. In 2015, a partial inspection of the source code revealed different technical miscodes which reduced the probability of a DNA sample matching the defendant.[28]

Scientific Miscode A scientific miscode refers to algorithms containing dubious scientific methods or value-judgments that affect the *validity*[29] of the output. A scientific miscode is not an unintentional error, nor is it necessarily a wrong method. However, the soundness and fairness of the assumption on which the program is based could be questionable since, for example, it might be an obsolete scientific method, it might be irrelevant to the issues sought to be proven,[30] or it might have indirect discriminatory effects. Some examples include the following:

COMPAS [Decision-Support Algorithm]
COMPAS is an automated risk-assessment software used in US courts to assess a criminal defendant's likelihood of becoming a recidivist. An independent analysis by ProPublica found that black defendants were far more likely than white defendants to be incorrectly judged at higher risk of recidivism, while white defendants were more likely than black defendants to be incorrectly assessed as low risk.[31]

---

[27] Van Buskirk and Liu (2006), p. 19.

[28] Chessman (2017) and Health Support Queensland and Forensic and Scientific Services (2016).

[29] The quality of being logically or factually sound.

[30] Jenkins (1975).

[31] Angwin and Larson (2016).

Intoxilyzer 500 [Algorithm-Based Evidence]
While breath alcohol content can be used to estimate blood alcohol content, the conversion factor for breath to blood (BAC) can vary from person to person. This particular device operated on the basis of Henri's Law and used a partition ratio of 2100:1. However, not every person has this ratio. "A wrong partition ratio can cause significantly erroneous results, as people with a lower ratio may have their BAC overestimated, and those with a higher ratio may have their BAC underestimated. Research has shown that many other factors can also influence the results of a breathalyzer. A person's gender and race are among the intrinsic factors that can impact their partition ratio".[32]

Legal Miscode A legal miscode refers to algorithms containing a misguided or even unlawful design that affects the *legality*[33] of the output. A legal miscode can be generated, for instance, from input data illegally obtained (*e.g.* in breach of privacy law) or from an unlawful criterion used for reaching a decision.[34] Furthermore, algorithms may be embedded in political and normative values that *legislators* are required to choose. This also gives rise to a legitimacy problem, as this kind of algorithm has the potential to distort the values underlying laws and policies which society has collectively agreed upon through democratic processes.[35] One could consider the following examples:

Malware Instruments for Investigations [Algorithm-Based Evidence]
Malware is a malicious software that is able to remotely control a digital device. It can have access to every functionality of the intercepted device, such as the microphone, camera and internal storage. The use of malware in criminal investigations as means of obtaining digital evidence has recently been regulated in Germany and in Italy. In the Italian code of criminal procedure, only microphone activation is allowed, it is necessary to have a court order and it is possible to use malware only in investigations into the most severe of crimes. Once again, the software may contain an "illegal design" which allows legally prohibited operations to be performed (*e.g.* the activation of the camera or copying of internal storage). This is precisely what happened with the Exodus, a malware that was sold to the Italian government by a private company for use in criminal investigations.[36] Due to an error in its

(continued)

[32]Liebow (2010).
[33]The quality or state of being in accordance with the law.
[34]See §4.2 as an example.
[35]Green (2018).
[36]Security without Borders (2019).

code, Exodus lacked a mechanism for validating the target. Consequently, it did not properly check whether the phone on which it had installed itself was intended to be legally intercepted or not.

Volkswagen Case [Algorithm-Based Evidence]
The most famous and clear example of legal miscode is the "Dieselgate scandal". In 2015, Volkswagen was embroiled in the emission-cheating scandal. Federal pollution regulators discovered that the company had installed illegal defeat devices in its diesel vehicles. This software was designed to make the vehicles meet the pollution standards when monitored in tests, whereas during actual use their emissions exceeded the legal limits in US.

3 Opacity in Algorithms Keeps Miscodes Hidden

Section 1 showed how algorithms may potentially contain miscodes. Consequently, the reliability of algorithms should be verified rather than taken for granted. Hence, the relevant question is: how can miscodes be detected? It is necessary to know how the algorithmic system works.

However, algorithms are usually concealed in secrecy and opacity that limit insight into their functioning. Without comprehending how the output has been generated,[37] it is not possible to test its accuracy, validity and legality (see Fig. 2).

Algorithmic opacity can be of different kinds. According to Burrell, opacity can stem from intentional corporate or state secrecy, technical illiteracy or from the lack of interpretability.[38] Pasquale, on the other hand, distinguishes between three types

Fig. 2 Algorithmic opacity

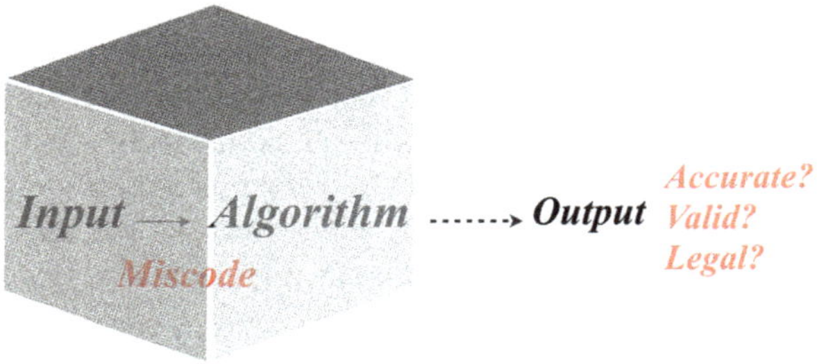

[37]Security without Borders (2019), Diakopoulos (2014) and Eslami et al. (2019). According to Pasquale, opacity is a blanket term for "remediable incomprehensibility".
[38]Burrell (2016).

of secrecy: legal, real and obfuscation.[39] Arguably, the notion of opacity not only refers to barriers between access and information, but it also concerns the issue of understandability of such information. In particular, this second aspect has become crucial after the rise of AI systems.[40] Traditional software, in fact, is based on a fixed and known model, whereas AI systems operate in environments that are at best partially known, even by the system designer.[41] In this case, understanding how the output has been generated is more of a challenge.[42]

A growing number of scholars from several disciplines are currently studying the issue of algorithmic opacity from different angles. The major trend is to consider "algorithmic transparency" (as opposed to opacity) a key-instrument for enhancing accountability of these systems,[43] for providing explanations to individuals affected by their use[44] and for limiting the adverse effects of the 'Black Box' on society.[45] Nevertheless, the concept of algorithmic transparency has also been criticized for its shortcomings. Particularly, Ananny and Crawford argue that transparency, in its literal significance as "being able to see a system", does not equate to understand algorithms.[46] We then "see without knowing".[47] Furthermore, transparency can even have undesirable consequences, such as allowing strategic gaming of the system.[48] In the light of the limitations of transparency, a second strand in the literature refers to the concepts of explanation and interpretability.[49]

However, there is still no conceptual clarity and uniformity of these notions among scholars. As Mittelstadt and Wachter underline, researchers in explainable AI (XAI) agree on the importance of providing explanations, but very few stop to check what they are agreeing to.[50] In the authors' view, these concepts can serve

---

[39]According to the author, real secrecy establishes a barrier between hidden content and unauthorized access to it. Legal secrecy obliges those privies to certain information to keep it secret. Obfuscation involves deliberate attempt at concealment when secrecy has been compromised. See Pasquale (2015).

[40]With regard to the General Data Protection Regulation, Malgieri and Comandé (2017) refers to the concept of "legibility" that is the capability of individuals to autonomously understand the logic, the significance and the envisaged consequences of an algorithmic decision-making.

[41]While datasets may be extremely large but possible to comprehend and code may be written with clarity, the interplay between the two in the mechanism of the algorithm is what yields the complexity (and thus opacity). See Burrell (2016).

[42]See in particular Beillevaire (2017).

[43]Kroll et al. (2016), Ananny and Crawford (2018), De Laat (2017) and Diakopoulos (2014).

[44]Kim and Routledge (2018), Edwards and Veale (2018), Doshi-Velez et al. (2017) and Burrell (2016).

[45]Pasquale (2015), Citron and Pasquale (2014), O'Neil (2016) and Eubanks (2017).

[46]Ananny and Crawford (2018).

[47]Ananny and Crawford (2018).

[48]Lepri et al. (2018).

[49]Lipton (2017), Lepri et al. (2018), Mittelstadt et al. (2019), Kim and Routledge (2018), Doshi-Velez et al. (2017) and Selbst and Powles (2017).

[50]Mittelstadt et al. (2019).

many audiences (e.g. experts or individual affected by a system's output) and functions (e.g. compliance with legislation, improving the performance of a system, enhancing accountability).[51] Likewise, these concepts should be further specified according to (but not limited to) the following questions: (1) who needs transparency/explanation and (2) why? The interest of this chapter is solely in ways of providing insight into algorithmic systems (1) to the defence (2) for the purpose of contestability of algorithm-based evidence and decisions in criminal proceedings. This will be further clarified in the following section through an analysis of the European fair trial rights.

4 Algorithmic Opacity Impacts Fair Trial Rights

Algorithmic opacity keeps miscodes hidden. Specifically, it is neither possible to detect technical errors, to know the method on which the algorithm is based, nor to verify the legality of its outputs. In criminal proceedings, the output of an algorithm constitutes a *statement* that is used for evidentiary purposes (*e.g.* for GPS evidence the statement that "the 4th of March the defendant was driving in Paris from Paris Gare du Nord and arrived at Rue des Abesses at 11.23 p.m.") or for supporting decision-making (*e.g.* for a risk-assessment software the statement that "the defendant is at low risk of recommitting a serious offence in the next 3 years"). However, without comprehending how the output has been generated it is not possible to contest these statements.[52] By preventing contestability, algorithmic opacity represents a significant threat to fair trial rights.

In this regard, the notion of "fair trial" provided by Article 6 of the European Convention of Human Rights will be adopted. The right to a fair trial has a position of pre-eminence in the European Convention of Human Rights[53] (hereafter ECHR). The key principle governing the application of this Article is the concept of "fairness", that is interpreted by the European Court of Human Rights (hereafter the Court) as providing only a procedural and not a substantive guarantee.[54] According

[51] See also in this sense: Mittelstadt et al. (2019).

[52] In the Loomis case the defense argued that the Court "should grant certiorari review to consider the due process implications of the increasingly common practice for using proprietary risk assessment instruments in sentencing defendants because the refusal of the companies producing those instruments to disclose significant information deprives defendants of the ability to assess the accuracy and validity of the information being used" and that "a defendant's right to due process is violated when he or she is denied access to information used at sentencing". See *State v. Loomis, No. 2015AP157-CR (Wis. Ct. App. Sept. 17, 2015), 2015 WL 1724741* (n 1).

[53] Harris et al. (2014), p. 370.

[54] See in particular: Harris et al. (2014), p. 371.

to the "fourth instance" doctrine, in fact, the Court is not in the position to deal with "errors of fact or law allegedly committed by a national court, unless and in so far as they may have infringed rights and freedoms protected by the Convention".[55] The Court's primary concern is to evaluate the *overall fairness* of the criminal proceeding.[56]

Article 6 ECHR is composed of three paragraphs that set general and specific procedural guarantees granted to the accused person in criminal cases. The general guarantees set in Article 6(1) ECHR are the right of access to a court, the right to an independent and impartial tribunal established by law, the right to a public hearing and to the public pronouncement of judgment and the right to be tried within a reasonable time. Moreover, a number of rights have been added through the medium of the right to a "fair hearing", such as such as the right to adversarial proceedings, to equality of arms between the parties, to access to the case file, to a reasoned judgment, to evidence, and the principle of legal certainty. Article 6(2) ECHR establishes the presumption of innocence of the accused, while (3) grants the accused person a non-exhaustive list of additional rights.

Article 6 ECHR applies when a "criminal charge" is being determined. This concept has an autonomous meaning in the Convention. In the determination of a "criminal offence", the starting-point for the assessment is based on the criteria outlined in *Engel and Others v. the Netherlands*,[57] that are as follow: (1) classification in domestic law (2) nature of the offence and (3) severity of the penalty that the person concerned risks incurring. According to the Court's case-law, a "charge" exists from the moment that an individual is officially notified by the competent authority of an allegation that he has committed a criminal offence, or from the point at which his situation has been substantially affected by actions taken by the authorities as a result of a suspicion against him.[58] As a result, the guarantees of the fair trial are applicable throughout the entirety of proceedings, including the pre-trial stage,[59] the sentencing process,[60] appeal proceedings against conviction or sentence that are in fact provided[61] and, similarly, the execution of final judgments in criminal cases.[62] Consequently, within this timeline, the use of algorithms for

---

[55]Garcia Ruiz v. Spain [GC], no. 30544/94, §28, ECHR 1999-I.

[56]See, among many others, Taxquet v. Belgium [GC], no. 926/05, §84, ECHR 2010; Schatschaschwili v. Germany [GC], no. 9154/10, §101, ECHR 2015; Ibrahim and Others v. the United Kingdom [GC], no. 50541/08, 50571/08, 50573/08 and 40351/09, §250, ECHR 2016.

[57]Engel and Others v. the Netherlands, 8 June 1976, §§ 82–83, Series A no. 22.

[58]See Deweer v. Belgium, 27 February 1980, §§ 42–46, Series A no. 35; Eckle v. Germany, 15 July 1982, § 73, Series A no. 51; McFarlane v. Ireland [GC], no. 31333/06, §143, ECHR 2010 and Ibrahim and Others v. the United Kingdom [GC], no. 50541/08, 50571/08, 50573/08 and 40351/09, §249, ECHR 2016.

[59]Dvorski v. Croatia [GC], no. 25703/11, §76 ECHR 2015.

[60]Phillips v. the United Kingdom, no. 41087/98, §39, ECHR 2001-VII.

[61]Eckle v. Germany, 15 July 1982, § 73, Series A no. 51.

[62]Harris et al. (2014), p. 378.

evidentiary purposes or for support decision-making must be compatible with the guarantees of a "fair trial" set in Article 6 ECHR.

With regard to algorithm-based evidence, the right to confrontation is particularly relevant. For the purpose of a fair trial, the defendant must have the possibility of confronting evidence against him.[63] Specifically, Article 6(3)(d) ECHR enshrines the principle that, before an accused can be convicted, "all evidence against him must normally be produced in his presence at a public hearing with a view to adversarial argument".[64] In the light of the autonomous and broad interpretation of the term 'witness' adopted by the Court, this provision applies to algorithm-based evidence when such evidence is (1) used at trial and appeal proceedings and (2) when it can serve to a material degree as the basis for a conviction.[65] According to the leading case *Al-Khawaja and Tahery v UK*, the right to confrontation not only requires that defendants should be in a position to challenge the *probity* and *credibility* of their evidence, but that they should also be able to test their *truthfulness* and *reliability.*[66] As showed above, algorithmic opacity is an obstacle to this test. In order to achieve the purpose of the confrontation, in fact, defendants needs to have insight on the algorithmic system. Oftentimes, this cannot be easily achieved as they might lack access to information and/or resources to understand it.

> Insofar as individuals in a legal process are unable to understand and contest, even with the help of legal counsel, complex algorithmic systems used to process evidence alleged to relate to them, there is a significant threat to due process rights.[67]

With regard to decision-support algorithms, the main legal issue relates to the lack of motivation and justification of such decisions. Article 6(1) ECHR requires reasoning of judicial decisions. According to established case-law, judgments of court and tribunals should adequately state the reasons on which they are based.[68] Reasoning, in fact, is instrumental to demonstrating to the parties that they have been heard, to preserve the rights of the defense and to allow and applicant to usefully exercise the right to appeal.[69] Moreover, they oblige judges to base their reasoning on objective arguments. This must be valid also when an algorithms is used for supporting judicial decisions.[70] Consequently, the "reasons" for the suggested decision must be comprehensible to (1) the parties, so as to be able to challenge

---

[63]Arslan (2018).

[64]Al-Khawaja and Tahery v UK [GC], no. 26766/05 and 22228/06, §118, ECHR 2011.

[65]For instance, in the case Georgios Papageorgiou v. Greece, the Court examined under Article 6§3 (d) ECHR the issue of access to the original documents and computer files relevant to the criminal accusations against the applicant.

[66]Al-Khawaja and Tahery v UK [GC], no. 26766/05 and 22228/06, §127, ECHR 2011.

[67]The Law Society of England and Wales (2019).

[68]Moreira Ferreira v Portugal [GC], no. 19867/12, §84, ECHR 2017.

[69]Hadjianastassiou v Greece, no. 12945/87, ECHR 1992.

[70]Under the rule of law, such decisions must also be comprehensible for those subject to the decision, and they must be justifiable in the sense of satisfying the legal norms that allow for the decision. Hildebrandt (2018).

the decision, and to (2) courts, so that they can consciously decide whether follow or digress from the suggestion. When opacity conceals the grounds on which the decision is based, defendants cannot exercise the right appeal.

To sum up, algorithmic opacity impacts the adversarial principle when the parties cannot contradict the opponent's allegations;[71] the equality of arms principle when it creates a knowledge asymmetry between parties;[72] the right to confrontation, when algorithms cannot be examined by the defence,[73] and the right to have a reasoned judicial decision when algorithms do not explain or justify how a particular decision has been reached. Henceforth, fair trial rights require insight into algorithmic systems. Although the algorithm-based evidence will be evaluated by the judge, and the decision-support algorithm will not totally remove human discretion, the defence must have the opportunity to access, understand and therefore *challenge* algorithm-based-evidence and decisions. As Hildebrandt argues:

> Contestability is at the heart of the rule of law, notably its procedural core. [. . .]The centrality of the adversarial procedure is meant to contribute to better decision-making (both in administration and in court), and this is precisely why the integration of ALI [Artificial Legal Intelligence] requires a new hermeneutics and a new adversarial design process.[74]

It is the role of procedural justice to ensure contestability and relieve the burden that opacity places on the defense asking for disclosure of information and seeking for expert knowledge to determine the significance of such information.[75] In this sense, many procedural questions should be carefully addressed to examine the practicability of potential solutions[76] and how they can be hinged in the criminal justice system. The criminal process in fact has two main goals: on the one hand the prosecution of crimes and the security of individuals and society in general, on the other hand the protection of individual rights. According to the Packer's model, balancing these two goals is where legitimacy lies.[77] Conversely, the goal of providing algorithmic contestability as instrumental for the safeguard of fair trial rights must be balanced with the other legitimate purposes of the criminal process, such as the protection of other individual rights,[78] the reasonable time of the proceeding and the need of legal certainty.

---

[71] See in this sense Chessman (2017), Imwinkelried (2016), Roth (2017) and Green (2018).

[72] See in particular Pagallo and Quattrocolo (2018).

[73] See in particular Celentino (2016), Neville (2011), Sites (2014), Chessman (2017) and Roth (2017).

[74] Hildebrandt (2018).

[75] See also Ananny and Crawford (2018).

[76] Such as the source code disclosure, independent validation, certificates, reverse engineering etc.

[77] Packer (1968).

[78] Secrecy in algorithmic design can be instrumental to the protection of the *ius excludendi alios*, such as proprietary protection, state or corporate secrecy or privacy rights.

5 The Harm Assessment Risk Tool (HART)

5.1 Introducing HART

The previous section concluded that algorithmic opacity challenges the core feature of fair trial rights. In this section, a concrete case-study is presented: the Harm Assessment Risk Tool (HART), an intelligent decision-support algorithm.

HART is one of the first software to be used by police forces in the UK for supporting custody decisions. Specifically, it is a machine learning tool that utilises random forest methodology to aid decision-making by forecasting three risk levels: high risk of committing a serious offence in the next two years; moderate risk of committing a non-serious offence in the next two years; low risk of commit any offences over the next two years. Individuals who are classified as moderate risk are eligible for the Constabulary's Checkpoint rehabilitation programme, that is thought to reduce offending and future harm in the community.[79]

HART uses past cases where the assessments have already taken place as a model of what will take place in the future.[80] For this purpose, the software was trained using approximately 104,000 custody events over a five year period within the systems of Durham Constabulary.[81] HART only suggests a particular risk-assessment, "recognized by the advisory nature of the algorithm".[82] However, even when decisions are made by both humans and technologies, it is still crucial to identify who is responsible for them[83] and to what extent police officers are influenced by the software outcomes.[84] According to Hildebrandt (2018):

> Even in the case of decision-support instead of decision-making, human intervention becomes somewhat illusionary, because those who decide often do not understand the 'reasons' for the proposed decision. This induces compliance with the algorithms, as they are often presented as 'outperforming' human expertise.

5.2 HART May Contain Miscodes

HART might hide all the three types of miscodes as classified *supra* in Sect. 1. Despite the little amount of information available to the public about the inner

[79]Oswald et al. (2018).

[80]Oswald et al. (2018).

[81]Data from other police force areas or indeed from national systems are not included.

[82]Oswald et al. (2018).

[83]This problem has been strongly highlighted by the Council of Europe's MSI-NET, 'Algorithms and Human Rights' (2018).

[84]In the HART context "will custody officers delegate responsibility to the algorithm because if they go against it they will undoubtedly face questions from higher up?", should they give any justification in case they contradict the software decision? See: Bridge and Swerling (2017).

working of this software, the documents of the Inquiry of the Science and Technology Committee and the reports made by Big Brother Watch have highlighted several concerns.[85]

As to scientific miscodes, the validity of its results should be questioned, in particular with regard to the predictor values used by the software. In order to reach a risk-assessment forecast, HART uses 34 different kinds of predictors, such as age, gender, and two forms of residential postcode, all combined with the offender's history of criminal behaviour. The use of postcode data seems to be problematic, as it can perpetuate indirect discriminatory bias towards geographical areas marked by community deprivation.[86] This problem was also underlined in the written evidence submitted by Durham Constabulary within the inquiry of the Science and Technological Committee of the House of Commons:

> There is an obvious concern with amplifying (or at least perpetuating) biases in models of this nature as the algorithms are built using historical outcome data. The model makes its predictions about the future based on offenders who have been arrested in the past. One could argue that, if the police target a deprived, high density neighbourhood/urban postcode, then more people from this area will come to police attention and be arrested than those living in an affluent, low density neighbourhood/rural postcode that does not see much police activity.

The risk is to create "a kind of feedback loop that may perpetuate or amplify existing patterns of offending".[87] Concerns have also been expressed in regard to other characteristics used in HART as potential sources of bias, especially where they serve as proxies for race or gender.[88] Big Brother Watch[89] also raised a concern about the Durham algorithm due to its inclusion of an Experian algorithm which could be discriminatory, since it characterizes people using metrics such as postcode, family composition and occupation.[90] The same criticism has been raised in a recent Report by The Law Society.[91]

As to the legal miscode, there must be considerable doubts over the legality of the use of certain predictor variables, so far as they could be prohibited as a criterion for

[85]House of Commons Science and Technology Committee (2018) and Big Brother Watch (2018).

[86]Using home address has been criticised by O'Neil (2016). HART's "postcode lottery" was also reported by Big Brother Watch (2018).

[87]Oswald et al. (2018).

[88]Professor Amoore explained how in the US, where algorithms had been used to predict the outcome in criminal trials, "even where race as a category was removed from the input data, the algorithm still learned characteristics, or attributes, that we might say are in breach of the Equality Act, because they use what we could call proxies for race. They learn patterns in past patterns of crime or they learn patterns in postcodes, for example". See the report House of Commons Science and Technology Committee (2018).

[89]Big Brother Watch is independent non-profit organisation leading the protection of privacy and civil liberties in the UK. For more information see https://bigbrotherwatch.org.uk/about/who-we-are/.

[90]Big Brother Watch (2018).

[91]The Law Society of England and Wales (2019).

risk-assessment. A clear example, provided by Oswald et al., is the inclusion of spent convictions in an algorithmic assessment of risk.[92] Under Section 4 of the Rehabilitation of Offenders Act 1974, a rehabilitated person shall be treated "for all purposes in law" as a person who has not committed the particular offence. As a consequence, spent convictions should not be a valid criterion in the risk-assessment of a rehabilitated person. Additionally, "data brokers such as Experian have been accused of acting illegally in their amassing of these datasets, and are subject to several high profile complaints from organizations such as Privacy International. For police forces to reuse such datasets when there is doubt about their lawful basis for existing seems problematic".[93]

Another issue of concern relates to the rate of errors of this software. Unsurprisingly, "as with any forecasting effort, HART inevitably produces errors".[94] However, on the basis of a value-judgement built into the algorithms, it gives different weights to different types of errors. As the Durham Constabulary explained:

> HART is able to forecast with 98% accuracy that for those forecast as low risk the very dangerous error will not occur. In order to do this the model will over predict cautious errors.

In essence, in order to reduce as far as possible the possibility of a high risk false negative (offender falsely predicted at low risk), which has been considered as the most dangerous error, the software overpredicts high risk false positives (offender falsely predicted at high risk) as compensation. As a consequence, high risk false positives have been deliberately made more likely to occur:[95] a low risk individual will be more likely to be kept in custody than a high risk individual is likely to be released. Indeed, the relevant question is: who made this choice? Although "these decisions may appear benign, they embed political values and causal assumptions within risk assessments and frame sentencing".[96] This choice actually involves a normative and a political judgment about the trade-off between reducing incarceration (and thus considering false positive as dangerous errors) and reducing crime (and thus considering false negative as dangerous errors).[97] These political design choices should never be outsourced.[98]

---

[92]Oswald et al. (2018).

[93]The Law Society of England and Wales (2019).

[94]'Written Evidence Submitted by Durham Constabulary' in House of Commons Science and Technology Committee (2018).

[95]Oswald et al. (2018).

[96]Green (2018).

[97]Green (2018).

[98]The Law Society of England and Wales (2019).

5.3 *Opacity in HART Keeps Miscodes Hidden*

Although the police force created a framework (called Algo-care) for assessing when these tools should be used by police, the opacity of HART remains hard to circumvent. Even if complete openness were granted, there would still be a problem in understanding the information, due to the 'intelligent' nature of this software.[99]

The HART model contains over 4.2 million decision points, all of which are highly interdependent on the ones that precede them within the tree structure and "unlike earlier methods of forecasting, it is simply not the case that a given input on a single predictor has an inflexible and inescapable impact on the forecasted outcome. Simply residing in a given post code has no direct impact on the result, but must instead be combined with all of the other predictors in thousands of different ways before a final forecasted conclusion is reached. It is therefore the combination of variables, and not the variables in isolation, that produces the risk level".[100] As explained by Oswald et al. (2018):

> These details could be made freely available to the public, but would require a huge amount of time and effort to fully understand.

5.4 *HART May Impact Fair Trial Rights*

Individuals affected by the use of HART should be able to enquire why they have been denied access to a rehabilitation programme.[101] Yet, the opacity of the system hinders insight on how the individual level of risk had been assessed.

It is important to underline that the Durham Constabulary is trying to address ethical and legal concerns in different ways. Firstly, in collaboration with the University of Winchester, they have developed an ethical framework called Algo-care, that seeks to provide guidance to police forces considering the use of algorithmic forecasting. The basic goal of this policy-document is move towards algorithms that can be considered lawful, accurate, challengeable, responsible and explainable.[102] In addition to this, especially in the context of the inquiry before the Science and Technological Committee of the House of Commons, they seemed willing to share as much information as possible and keen to have a dialogue on the main issues that arise from the use of HART. Including ethics in the design of the software will be crucial for enhancing fairness in its use. Nevertheless, this must be accompanied by *ex-post* mechanisms of scrutiny by the defence, providing insight into the system.

---

[99]See supra §2 and Oswald et al. (2018).

[100]'Written Evidence Submitted by Durham Constabulary' in House of Commons Science and Technology Committee (2018).

[101]See in particular the recommendation of The Law Society of England and Wales (2019).

[102]Burgess (2018).

At the moment, the information necessary to make a proper assessment over the actual impact of HART on fair trial rights is not sufficient. Sheena Urwin, Head of Criminal Justice of Durham Constabulary, had clarified in the written evidence submitted in February 2018 that they are:

> unable to publish any further detail as insufficient time has elapsed. It is our intention to publish research as and when it becomes available. I do appreciate there is a desire to see as much information as possible about HART, we too are keen to see the outcome of the research however we must wait for sufficient time to elapse.

Furthermore, the specific use of this tool and its procedural scope is still unclear from the information available. On the occasion of the inquiry before the Scientific and Technological Committee, Urwin affirmed that HART is only used to identify who is eligible for the Check-point rehabilitation program:[103]

> To be clear, the risk of reoffending does not determine whether or not someone is arrested, nor does it determine whether someone will be detained in custody for an extra 24hrs

However, in the second validation study conducted on HART, the authors (including Urwin) stated that:

> It is anticipated at the date of writing that the model will be live in late-2017, and that its use will *expand beyond the current experimental Check- point treatment program*, with the forecasts influencing *all of the many other decisions* that are made in the wake of bringing a suspected offender into police custody.[104]

If HART will be actually used for other purposes, including custodial decisions with a higher impact on individual's freedom, it will be crucial to safeguard the rights of the defence. Solutions can be envisioned taking into account the different types of miscodes. As shown in Sect. 5.2, miscodes can also relate to the legality of some predictor variables. For this case, a way to ensure contestability, although not sufficient, would be providing the defense with a detailed and certified documentation listing the predictor values embedded in the system. Moreover, the weight or the interaction between this variables should also be explained, especially with regard to those categories that can serve as proxies for race and gender, assuming this is possible due to the intelligent nature of the software.[105] The availability of the training data should also be considered, taking into account, however, that compelling extensive access to information might worsen, rather than improve, their actual comprehensibility and it may also impact privacy rights of individuals. Additionally, certificates attesting accurate and updated tool testing of the software would be indeed helpful for preventing the likelihood of technical miscodes.

---

[103]The program identifies why an individual adult has offended, along with the best interventions and services to support the individual in turning away from crime. In order to pursue this strategy, Checkpoint must identify those at an appropriate risk of reoffending.

[104]Oswald et al. (2018).

[105]An emerging idea is the use of reverse engineering, as an alternative to disclosure of source code, to provide ex-post explanations. See in this sense Diakopoulos (2014). Another method is via quantitative input influence. See Datta et al. (2016).

Surely, these suggestions are not sufficient to cope with this issue in its entirety.[106] In my view, one of the most challenging aspects remains the issue of understandability of algorithmic systems, that is hampered by the complexity of AI and by the lack of technical expertise (or resources to afford it!) of the defense.

Other proposed solutions in the literature include providing extensive training, auditing program,[107] external validation and/or attesting the configuration of the software,[108] access to source code[109] or, alternatively, written answers about how the machine works, the presence of a defence expert at the time of testing to discourage and unearth case-specific input errors.[110] In the trial stage, cross-examination of who used the software, access to the necessary documentation regarding how the machine is functioning, through testimony of a defence expert witness[111] and testimony from the programmers of machines performing forensic analyses.[112] Nevertheless, the practical solutions proposed must be assessed taking into account the specific right at stake and the stage of the proceedings in which these tools are used. However, if the opacity surrounding HART will be such as to prevent the contestability of the decision based on it, its use should not be admissible. In the trade-off between opportunities and risks, the protection of fair trial rights has to prevail. Technological progress can be beneficial in all its aspects, if it goes hand-in-hand with the protection of fundamental rights.

6 Conclusion

The use of algorithms in criminal justice is widespread and gives rise to concerns about the trade-off between its benefits and its dangers.[113] The risk of transforming the criminal justice system into an "automatic system of machinery",[114] in which the role of humans is proportional to what the machine cannot (yet) do, is real. Hence, an appropriate model of algorithmic governance in criminal procedure is needed for safeguarding, in particular, the core features of fair trial and due process.

---

[106] I want to underline here that this chapter illustrates the findings of my first year of PhD project lasting four years.

[107] Citron (2016).

[108] Bratus et al. (2010).

[109] Chessman (2017).

[110] Roth (2017).

[111] Watson (2015).

[112] Neville (2011).

[113] The Law Society of England and Wales (2019).

[114] In the so-called "Fragment on Machines", Marx describes this system of numerous mechanical and intellectual organs that he believes will radically change the relationship between labour, production and means of production and in which the worker will cast merely as "a conscious linkage" of the automaton. See K. Marx, Grundrisse: foundations of the critique of political economy.

The aim of this chapter was to show that opaque algorithmic evidence and decision-support algorithms poses significant challenges to the right of the defence. By hindering insight into algorithmic systems, opacity prevents contestability of algorithmic-based evidence and decisions. Bearing in mind the importance of these new legal issue, it is crucial to conceptualize effective remedies against the adverse effect of algorithmic opacity.

Specifically in the context of criminal proceedings, any potential technical solutions must be hinged in the criminal justice system and shaped according to the requirements of the rights involved. In this regard, the categories that are proposed in this chapter are thought to be helpful for understanding the problems that arise from the use of black boxes in criminal law enforcement. In the author's view, these issues should be addressed taking into account the different types of miscodes, the specific stage of the criminal proceeding in which algorithms are used and the rights involved. Conclusively, the legitimate use of algorithms in criminal proceeding strictly depends on whether contestability of their outputs can be *de facto* and *de jure* guaranteed.

References

Ananny M, Crawford K (2018) Seeing without knowing: limitations of the transparency ideal and its application to algorithmic accountability. New Media Soc 20:973

Angwin J, Larson J (2016) How we analyzed the COMPAS recidivism algorithm. ProPublica, 23 May 2016. https://www.propublica.org/article/how-we-analyzed-the-compas-recidivism-algorithm. Accessed 27 Oct 2018

Arslan M (2018) The right to examination of prosecution witnesses. Zeitschrift für Internationale Strafrechtsdogmatik 6:218

Beillevaire M (2017) Inside the black box: how to explain individual predictions of a machine learning model. Master thesis, KTH Royal Institute of Technology School of Electrical Engineering and Computer Science

Big Brother Watch (2018) A closer look at Experian big data and artificial intelligence in Durham police. Big Brother Watch Blog, 6 April 2018. https://bigbrotherwatch.org.uk/2018/04/a-closer-look-at-experian-big-data-and-artificial-intelligence-in-durham-police/. Accessed 11 Dec 2019

Bratus et al (2010) Software on the witness stand: what should it take for us to trust it. In: Acquisti A et al (eds) Trust and trustworthy computing. Springer, Berlin, Heidelberg

Bridge M, Swerling G (2017) Bail or jail? App helps police make decision about suspect. The Times, 11 May 2017. https://www.thetimes.co.uk/article/bail-or-jail-app-helps-police-make-decision-about-suspect-kv766zjc9. Accessed 24 Jan 2019

Burgess M (2018) UK police are using AI to inform custodial decisions but it could be discriminating against the poor. Wired. https://www.wired.co.uk/article/police-ai-uk-durham-hart-check point-algorithm-edit

Burrell J (2016) How the machine "thinks": understanding opacity in machine learning algorithms. Big Data Soc 3. https://doi.org/10.1177/2053951715622512

Celentino JC (2016) Face-to-face with facial recognition evidence: admissibility under the post-crawford confrontation clause. Mich Law Rev 114(7):1317

Chessman C (2017) A source of error: computer code, criminal defendants, and the constitution. Calif Law Rev 105:179

Citron DK (2016) (Un)Fairness of risk scores in criminal sentencing. Forbes. https://www.forbes.com/sites/daniellecitron/2016/07/13/unfairness-of-risk-scores-in-criminal-sentencing/. Accessed 9 July 2019
Citron DK, Pasquale F (2014) The scored society: due process for automated predictions essay. Wash Law Rev 89:1
Cormen TH et al (2009) Introduction to algorithms. The MIT Press, Cambridge
Council of Europe's MSI-NET (2018) Algorithms and Human Rights. https://rm.coe.int/algorithms-and-human-rights-en-rev/16807956b5. Accessed 14 Apr 2020
Datta A et al (2016) Algorithmic transparency via quantitative input influence: theory and experiments with learning systems. 2016 IEEE Symposium on Security and Privacy (SP) 598
de Laat PB (2017) Algorithmic decision-making based on machine learning from big data: can transparency restore accountability? Philos Technol 31:525
Diakopoulos N (2014) Algorithmic-accountability: the investigation of black boxes. Tow Center for Digital Journalism
Dilek S et al (2015) Applications of artificial intelligence techniques to combating cyber crimes: a review. Int J Artif Intell Appl 6:21
Doshi-Velez F et al (2017) Accountability of AI under the law: the role of explanation. https://arxiv.org/abs/1711.01134
Edwards L, Veale M (2018) Enslaving the Algorithm: From a "Right to an explanation" to a "Right to better decisions"? IEEE Security & Privacy 16:46
Eslami M et al (2019) User attitudes towards algorithmic opacity and transparency in online reviewing platforms. In: CHI Conference on Human Factors in Computing Systems Proceedings (CHI 2019), May 4–9, Glasgow, Scotland UK. https://doi.org/10.1145/3290605.330072414
Eubanks V (2017) Automating inequality: how high-tech tools profile, police, and punish the poor. St Martin's Press, New York
Feldman B (2018) This is why Alexa is laughing at you. Intelligencer, 8 March 2018. http://nymag.com/intelligencer/2018/03/this-is-why-alexa-is-laughing-at-you.html. Accessed 19 Mar 2019
Freeman K (2016) Algorithmic injustice: how the Wisconsin Supreme Court failed to protect due process rights in State v. Loomis. N C J Law Technol 18:75
Gillespie T (2014) The relevance of algorithms. Media technologies: essays on communication, materiality, and society 167
Green B (2018) Fair risk assessments: a precarious approach for criminal justice reform. 5th Workshop on fairness, accountability, and transparency in machine learning
Harris DJ et al (2014) Law of the European Convention on Human Rights. Oxford University Press, Oxford
Health Support Queensland and Forensic and Scientific Services (2016) STRmix Miscoding Error. https://www.health.qld.gov.au/__data/assets/pdf_file/0029/633368/dohdl1617012.pdf. Accessed 13 Feb 2019
Hildebrandt M (2018) Algorithmic regulation and the rule of law. Philos Trans R Soc A. 37620170355
Horsman G (2019) Tool testing and reliability issues in the field of digital forensics. Digit Investig 28:163
Hotten R (2015) Volkswagen: the scandal explained. BBC News, 10 December 2015. https://www.bbc.com/news/business-34324772. Accessed 2 Apr 2019
House of Commons Science and Technology Committee (2018) Algorithms in decision-making. House of Commons 2018. Fourth Report of Session 2017–2019
Imwinkelried EJ (2016) Computer source code: a source of the growing controversy over the reliability of automated forensic techniques. DePaul Law Rev 66:97
Jenkins MM (1975) Computer-generated evidence specially prepared for use at trial law and technology symposium. Chicago-Kent Law Rev 52:600
Kim TW, Routledge B (2018) Algorithmic Transparency, a Right to Explanation and Trust. https://www.researchgate.net/publication/327160677_Informational_Privacy_A_Right_to_Explanation_and_Interpretable_AI
Kroll JA et al (2016) Accountable algorithms. Univ Pa Law Rev 165:633

Lepri B et al (2018) Fair, transparent, and accountable algorithmic decision-making processes. Philos Technol 31:611
Liebow D (2010) DWI Source Code Motions after Underdahl. Minn J Law Sci Technol 11:853
Lipton ZC (2017) The mythos of model interpretability. arXiv:1606.03490 [cs, stat]
Longo A (2019) Più di mille italiani intercettati sul cellulare, per errore, da un hacker di Stato. La Repubblica, 30 March 2019. https://www.repubblica.it/tecnologia/sicurezza/2019/03/30/news/molte_centinaia_di_italiani_intercettati_su_cellulare_per_errore_da_hacker_di_stato-222865990/. Accessed 2 Apr 2019
Maiberg E et al (2019) Researchers find Google Play Store apps were actually government malware. Motherboard. https://motherboard.vice.com/en_us/article/43z93g/hackers-hid-android-malware-in-google-play-store-exodus-esurv. Accessed 1 Apr 2019
Malgieri G, Comandé G (2017) Why a right to legibility of automated decision-making exists in the General Data Protection Regulation. Int Data Privacy Law 7:243
Marx K, Nicolaus M. *Grundrisse*: foundations of the critique of political economy (rough draft)
Mittelstadt B et al (2019) Explaining explanations in AI. Proceedings of the Conference on Fairness, Accountability, and Transparency - FAT* '19. ACM Press 2019
Neville K (2011) Programmers and forensic analyses: accusers under the confrontation clause. Duke Law Technol Rev 10:1
O'Neil C (2016) Weapons of math destruction: how big data increases inequality and threatens democracy. Crown, New York
Oswald M et al (2018) Algorithmic risk assessment policing models: lessons from the Durham HART Model and "Experimental" Proportionality. Inf Commun Technol Law 27:223
Packer HL (1968) The limits of the criminal sanction. Stanford UP; Oxford UP
Pagallo U, Quattrocolo S (2018) The impact of AI on criminal law, and its two fold procedures. In: Barfield W, Pagallo U (eds) Research handbook on the law of artificial intelligence. Edward Elgar, Cheltenham
Pasquale F (2015) The black box society: the secret algorithms that control money and information. Harvard University Press, Cambridge
Roth A (2016) Trial by machine. Georgetown Law J 1245
Roth A (2017) Machine testimony. Yale Law J 126(1):1972. https://www.yalelawjournal.org/article/machine-testimony. Accessed 12 Mar 2019
Sauer G (2017) A murder case tests Alexa's devotion to your privacy. Wired. https://www.wired.com/2017/02/murder-case-tests-alexas-devotion-privacy/. Accessed 19 Mar 2019
Security without Borders (2019) Exodus: new android spyware made in Italy. https://securitywithoutborders.org/blog/2019/03/29/exodus.html. Accessed 1 Apr 2019
Selbst AD, Powles J (2017) Meaningful information and the right to explanation. Int Data Privacy Law 7:233
Sites B (2014) Rise of the machines: machine-generated data and the confrontation clause. Columbia Sci Technol Law Rev 16:36. https://ssrn.com/abstract=2548701
The Law Society of England and Wales (2019) Algorithm use in the criminal justice system report. https://www.lawsociety.org.uk/support-services/research-trends/algorithm-use-in-the-criminal-justice-system-report/?fbclid=iwar00xj3zliogp71ykvnzlhuaxyxlrklc1qkefczkrijfbharotgn_-omr5u. Accessed 1 July 2019
Van Buskirk E, Liu VT (2006) Digital evidence: challenging the presumption of reliability. J Digit Forensic Pract 1:19
Vowles E, Story Carter J (2018) Your smart device could be used against you in Court. ABC News, 9 March 2018. https://www.abc.net.au/news/2018-03-09/your-google-home-or-fit-bit-could-be-used-against-you-in-court/9510368. Accessed 19 Mar 2018
Watson KE (2015) COBRA data and the right to confront technology against you. North Ky Law Rev 42:375

Microchipping Employees: Unlawful Monitoring Practice or a New Trend in the Workplace?

Seili Suder and Merle Erikson

Contents

Abstract A specific technology has been making its way into a growing number of workplaces. In addition to computer screen recording, video surveillance, keystroke monitoring, location tracking and social media monitoring, employers can now use microchips as an additional tool to monitor employee activity. These microchips give new abilities to employees and replace credit cards, keys, passwords and bracelets. Indeed, microchips allow employees to automatically open doors, trigger computers or printers and pay for purchases. Despite these useful qualities, microchips may bring forth several possible legal concerns, such as the possibility to covert and constant surveillance, profiling and digital discrimination. As this innovative technology appears destined to further challenge regulatory frameworks, this chapter aims to provide answers to the following research questions: is GDPR applicable in case of microchipping? If GDPR is applicable in these situations, then under what legislative grounds can microchipping be lawfully accommodated? Also, as each EU Member State could provide more specific rules under the GDPR to safeguard the employees' right of personal data, the chapter analyses whether more specific national rules are needed in order to protect microchipped employees.

S. Suder (✉) · M. Erikson
University of Tartu, Faculty of Law, Tartu, Estonia
e-mail: seili.suder@sm.ee

M. Ebers, M. Cantero Gamito (eds.), *Algorithmic Governance and Governance of Algorithms*, Data Science, Machine Intelligence, and Law 1,
https://doi.org/10.1007/978-3-030-50559-2_4

1 Introduction

I'm turning the Internet of Things into the Internet of Us. (Jowan Osterlund, the founder of Swedish company Biohax that specialises Kin injecting microchips)[1]

Today, new applications and smart devices enable employers to collect enormous quantities of employees' personal data and to do so within a reasonable time and with inexpensive means.[2] Amid this convergence of technology and workplace surveillance, there is a specific technology that has been making its way into a growing number of workplaces. In addition to computer screen recording, video surveillance, keystroke monitoring, location tracking using wearable technology and global positioning systems and social media monitoring, employers now have Radio Frequency Identification (RFID) microchips[3] as an additional tool to monitor employee activity.

Initially microchips, were used to track livestock and pet animals for their proper identification.[4] These chips can replace house and car keys, store train tickets and might be used in the future in place of a passports. Today, however, chips are being injected into human bodies worldwide for a variety of reasons, such as personal recreational use, research and medical applications.[5] The basic goal of a microchip is to automatically identify and track objects (in this case humans). These microchips give new abilities to employees and replace artefacts such as credit cards, keys, passwords and bracelets. Indeed, microchips allow users to automatically open doors, trigger computers or printers and pay for purchases.[6] Chips are linked to trivial activities, but as the cost of implementing RFID systems drops and the functionality of chips grows, organizations may choose to use this technology to track employee's productivity, improve security and reduce theft.[7]

In fact, the practice of microchipping has become a reality in workplaces. There are various examples of chips being implanted into employees in different parts of the world (USA, Mexico, Sweden, Belgium).[8] This rising new trend is also apparent in Estonia were presently few companies have taken the lead in microchipping their employees.[9]

[1]Glaser (2017).

[2]Ogriseg (2017).

[3]RFID devices are known as a technology for the remote identification. RFID uses tags (or transponders) with up to two kilobytes of data embedded in a microchip to locate, identify and transmit information on items (or people) carrying the chip. RFID does not require contact or line-of-sight for data capture. MHI (2019).

[4]Khan (2015).

[5]Rodney et al. (2008).

[6]Gauttier (2019).

[7]Office of the Privacy Commissioner of Canada (2008).

[8]Martins Esfola (2018).

[9]Kanarbik (2018).

The reason for the recent escalating use of chips lies primarily in advanced miniaturisation and in the constant price decline which makes the use of chips economically viable.[10] Although the exact number of microchipped employees in the world is currently unknown, it is clear that the figure is rising. Preliminary findings indicate that our technology saturated society is also becoming increasingly accepting of *insideable* technologies, e.g. technologies put under the skin.[11] With the help of commercialisation the chips could, however, become even more commonplace.[12] Just as the popularity of CCTV cameras rapidly grew into public acceptance, human microchipping may also become a normal practice one day. Considering the dominating infatuation with various technological devices and the sophistication of new dataveillance technologies, it is only inevitable to think that this field will continue to evolve and grow in popularity. In 2017, a Global Shapers Survey concluded that 44.3 percent of the young would not agree with embedding an implant under their skin or in their brain when faced with a possibility to increase one's capabilities.[13] Still, the members of the millennial generation, however, who are more used to body-tech may be more likely to embrace this new technology in the near future.

In addition to positive functions, chips can also be characterized by their continuous presence and lack of control of the employee. Due to these qualities, the use of microchips in a manner that is respectful of privacy is crucial to maintaining human dignity in the employment context.[14] In our opinion, the use of microchips may bring forth several data protection concerns, such as the possibility to covert and constant surveillance, profiling and digital discrimination s (e.g. more prejudiced treatment of employees without chips which may lead to their exclusion from different possibilities such as better working conditions).

The European Parliament has also acknowledged data protection concerns microchips could bring to the employees and published a study on possible issues arising from microchipping in the workplace. Still, the analysis of such problems was not far fetched.[15] In fact, increasing technology innovation appears destined to further challenge regulatory frameworks as personal data collection technologies can be expected to develop in the future.[16] For example, the growth potential of microchipping and privacy concerns have lead several states in the U.S. to proactively adopt legislation to contain the threat posed to employees.[17] Given the potential for increased workplace surveillance through the use of microchips, it is

[10]Thiesse (2007).

[11]Thiesse (2007) and Gauttier (2019).

[12]Roosendaal and Kosta (2008).

[13]World Economic Forum (2017).

[14]World Economic Forum (2017) and Office of the Privacy Commissioner of Canada (2008).

[15]Graveling et al. (2018).

[16]Tikkinen-Piri (2017).

[17]Schmidt and O'Connor (2018).

also time to analyse if the General Data Protection Regulation (GDPR)[18] is up to the challenge. The issue is even more relevant in the employment context as the GDPR does not provide specific guidelines for its provisions' implementation and has not given a set of uniform rules for employees' data protection.[19]

This chapter questions if the current EU data protection rules are adequate to deal with the challenges of microchipping and asks if we are in need of new rules to work alongside the advances of this technology. More specifically, the chapter aims to provide answers to the following research questions: Is GDPR applicable in case of microchipping? If GDPR is applicable in these situations, then under what legislative grounds can microchipping be lawfully accommodated? Also, as each EU Member State could provide more specific rules under the GDPR to safeguard the employees' right of personal data,[20] the paper analyses whether more specific national rules are needed in order to protect microchipped employees.

2 Microchips as a Source of Employee's Personal Data?

In this section we analyse if data received from employees' microchips triggers the applicability of GDPR. For GDPR to be applicable, employer's activities have to amount to the processing of employee's personal data.[21] Personal data in the GDPR means "any information relating to an identified or identifiable natural person".[22] Therefore, in order the GDPR to be applicable, microchips have to give out information that is linked to an employee and can, on its own or combined with other information, lead to the identification of that employee.[23]

In case of microchipping employees the question of personal identifiable data can be a crucial one, as due to different types of Radio Frequency Identification (RFID) systems, employer may or may not have access to personal data. We explain this by giving a short description of a RFID system.

The RFID system consists of two main components, a chip, which is implanted in the employee and the reader, which is used to read its data without contact. The role of the reader is to interrogate the chip and retrieve its stored data.[24] To do this the

[18]European Parliament and Council Regulation (EU) 2016/679 of 27 April 2016 on the protection of natural persons with regard to the processing of personal data and on the free movement of such data, and repealing Directive 95/46/EC (2016) OJ L119.

[19]Ibid. Ogriseg (2017).

[20]Ibid. European Parliament and Council Regulation (EU) 2016/679. Article 88.

[21]We did not analyse the notion of processing in this chapter as the definition for processing under Article 4(2) of GDPR is broad and encompasses every kind of operation possible on personal data. Regardless of the specific activity nor technology used, employer's activities fall under the notion of 'processing' described in the GDPR.

[22]Ibid. European Parliament and Council Regulation (EU) 2016/679. Article 4(1).

[23]Information Commissioner's Office (2012).

[24]Meints (ed) (2007).

reader broadcasts an electromagnetic signal that any RFID chip within range and operating on the same frequency will respond to. Drawing its power from this signal, the RFID chip will respond with an encrypted signal.[25] The RFID reader will decode this and pass the resulting information to the backend system. The backend system is where data is manipulated and stored, and forms the data resource for the system users.[26]

RFID systems may be built so that the employer does not retrieve any personal information about the employee. However, there are several ways that the data contained on a chip may become personal data. This possibility is also suggested in the GDPR that does not mention "RFID", but clearly states that "[n]atural persons may be associated with online identifiers provided by their devices, applications, tools and protocols, such as ... radio frequency identification tags".[27]

So how can an employer identify the employee carrying the microchip? First of all, the chip itself may contain information about the employee. In other words, the chip can function as an information carrier, bearing for instance the person's name and address, age or medical condition.[28] Furthermore, a photograph, a fingerprint or a retinal scan may also be included which would enable facial or biometric identification so as to enhance security.[29] When information about an employee is stored on the chip, it is beyond doubt that this information qualifies as personal data.

Secondly, but more commonly, the chip only contains a unique number.[30] This unique number is useless as a piece of data in isolation. However, the Court of Justice of the European Union (CJEU) has ruled that a dynamic IP address can be personal data.[31] So the question arises, whether a unique chip identifier can be regarded in a similar manner. In case of microchipping, the backend system typically incorporates a database allowing the matching of the chip's data to other stored information or storing of the chips details themselves.[32] The unique number becomes relevant if it can be linked to an employee and the employee is eventually identifiable with the help of data stored in a database.[33] In, such case, the information would be considered as personal data and GDPR would apply.

Even in situations where chips use anonymization techniques on the data collected, it is possible to reveal the identity of the person. For example, a backend system may record the occurrences of chip in Mary's room every day and quickly

---

[25]Meints (ed) (2007) and Graveling et al. (2018).

[26]Graveling et al. (2018) and Meints (ed) (2007).

[27]Graveling et al. (2018) and Meints (ed) (2007). European Parliament and Council Regulation (EU) 2016/679. Recital 30.

[28]Halperin et al. (2008).

[29]Halperin et al. (2008) and Graveling et al. (2018).

[30]Fischer-Hübner and Hedbom (eds) (2008).

[31]Patrick Breyer v Bundesrepublik Deutschland (2016) Judgment in Case C-582/14. Court of Justice of the European Union.

[32]Ibid. Meints (ed) (2007).

[33]Ibid. Roosendaal and Kosta (2008).

assume that the implanted person is Mary. Therefore, when chips are used for their ability to monitor location and if the implanted employee is performing her tasks in a specific workplace, it is also possible that, through the analysis of employee's movements, her identity would be obtained. Thus, again we are certain that the situation falls under the scope of the GDPR.

Furthermore, the employer also needs to think about the potential of using personal data from the chips in the future as the Article 29 Working Party[34] is convinced that the future technological possibilities have to be considered when classifying personal data.[35] This task is increasingly important in case of microchips due to potential enhanced capabilities which provide microchips with GPS tracking capability and voice activation software.[36]

In summary, microchipping enables the processing of employees' personal data in two ways. Either personal data are directly stored on the chip or can be indirectly linked to an employee by combining chip's unique number, with data stored in a backend system. If data received from microchips is linked to an employee who is identified or identifiable (today or in the future), it will become personal data and the processing of his/her data will fall under the scope of the GDPR.

3 Is Employee's Consent Enough to Justify Microchipping?

Personal data processing is allowed only under exhaustive list of legal bases mentioned in the GDPR. These bases are relevant because they represent a pre-condition for employees' personal data processing and render the processing unlawful if not directly relied upon. Data processing in an employment relationship is usually based on a contract.[37] However, contract can only work as a legal basis if data processing is necessary for the performance of the contract. In case of monitoring microchipped employees, this will probably not be the case as the use of chip's data is not indispensable for contractual obligations. Therefore, we exclude this ground as relevant for monitoring microchipped employees and proceed with other possible legal ground—consent.

To lawfully introduce microchips in the workplace, employer would need to obtain freely expressed consent from the employees. However, consent, in employment context, has repeatedly been criticized as it is likely to turn into an empty and ritual process.[38] In our opinion, employers who want to microchip their employees

---

[34]The Article 29 Working Party is the independent European working party that dealt with issues relating to the protection of privacy and personal data until 25 May 2018 (entry into application of the GDPR).

[35]Article 29 Working Party (2007).

[36]Rodriguez (2019).

[37]Rodriguez (2019). European Parliament and Council Regulation (EU) 2016/679. Article 6(1)b.

[38]Padi (2018).

will have a hard time to demonstrate that consent meets all the criteria laid down in the GDPR.

First of all, the consent 'should not be regarded as freely given if the data subject has no genuine or free choice or is unable to refuse or withdraw consent without detriment'.[39] Consent is said to be powerless in circumstances where a data subject requires a certain good,[40] as is the case in the context of employees. In fact, due to the imbalance of power that exists in the employment relationship employees are rarely able to exercise their free will.[41] Therefore, the employer should not be given the opportunity to use its position to pressure employees into agreeing to be microchipped.

Still, the GDPR does not prohibit the use of consent in the employment relationship.[42] In our opinion, not every employee is in a situation of clear imbalance of powers (e.g. employee working in a leading position or in a small business). We think, that in some cases consent may justify monitoring of microchipped employees. However, it is up to the employer to demonstrate that such monitoring was consented freely.

If consent is used as a legal basis for monitoring implanted employees, it should be restricted to situations where the employee is genuinely able to exercise free choice without any negative consequences.[43] Under GDPR, when assessing whether consent is freely given, utmost account shall be taken of whether, *inter alia,* the performance of a contract, is conditional on consent to the processing of personal data that is not necessary for the performance of that contract.[44] This provision contemplates an employment situation in which a job offer is made on a "take it or leave it" basis.[45] Therefore, employment should never be made contingent on a willingness to be implanted with a chip. However, refusing consent is often complicated in an employment relationship. For example, refusing a microchip can be difficult if an employer decides to grant access to a workplace only using chips or asks a potential candidate during a job interview if he will agree to microchipping. We are certain that in both of these cases employee is lead to consent to the use this new technology.

Sometimes, it is also difficult to refuse consent, i.e. refusing becomes somehow "abnormal".[46] For example, microchipped colleagues can set the trend for others, thus decreasing scepticism towards the technology[47] and generate a culture where

---

[39]Padi (2018). European Parliament and Council Regulation (EU) 2016/679. Recital 42.

[40]Belli et al. (2017).

[41]Article 29 Working Party (2017).

[42]Article 29 Data Protection Working Party (2017).

[43]Article 29 Data Protection Working Party (2017).

[44]Article 29 Data Protection Working Party (2017). European Parliament and Council Regulation (EU) 2016/679. Article 7(4).

[45]Mangan (2018).

[46]Poullet (2009).

[47]Poullet (2009) and Gauttier (2019).

employees are expected to get microchipped. In other circumstances, importance of consent can easily be by-passed. Employees may be unable to understand the consequences of consent decisions.[48] Consenting to microchipping may look interesting as employees may also benefit from the technology. For example, they can use services conveniently, explore the potential of the new technology and be part of the future that is promoted by the company.[49] These triggers of microchipping, e.g. the spread of individual cases, peer pressure, different affordances provided by the chip and the infrastructure in a workplace, have an impact on the quality of the consent given by the employee. As a consequence, employees may often be conditioned when giving consent to make a calculated and rational decision about microchips.

Most importantly, the voluntary implantation of microchips requires informed consent. Consent needs a clearly defined scope of action, i.e. the consenting employee needs to have the relevant information so that they would know what they are consenting to.[50] However, in case of microchipping an informed decision is hard to make. In our opinion, informed consent should include proper knowledge about the use of the data from the chip, the placement of readers and whether the data received from the chip is combined with other data. Employees should also know, that the design of the technology does not allow turning the chips off and the impossibility to shield ones' personal data that are on the devices from being read. Furthermore, attempts to meaningfully inform employees may also be inadequate. Research suggests, that highly technical, long and complex privacy notices or policies often fail to inform data subjects about the true nature of data processing practices.[51] Research shows that even if a company has guidelines and policies employees are not properly informed about the content of these documents. Furthermore, in the majority of cases, employees have simply forgotten the content or have never read it.[52] From these notices or policies it is often not clear to employees what data is collected, for what purposes, how the data is analysed, and which decisions result from the analyses.[53] Also, in a bid to avoid liability and take account of future data uses, it is foreseeable that employers may draft vague privacy policies to cover any unforeseen eventuality of processing.[54] We propose that, consent gathered by such privacy notices and policies is meaningless as broad and permissive information fails to genuinely inform the employee about all aspects and consequences of microchipping.

Luckily, the employee has the right to withdraw their consent at any time.[55] However, the possibility to withdraw one's consent in case of microchips seems

[48]Custers et al. (2013).

[49]Custers et al. (2013) and Gauttier (2019).

[50]Mitrou (2019).

[51]Special Eurobarometer 359 (2011).

[52]Suder and Siibak (2017).

[53]Solove (2013).

[54]Solove (2013) and Padi (2018).

[55]Ibid. European Parliament and Council Regulation (EU) 2016/679. Article 7(3).

difficult to achieve without exiting or limiting the 'service provided' by chips in the workplace.[56] Thus, employee must be able to withdraw their consent without detriment. In fact, we argue that an employer must pay for the surgery to remove the chip and guarantee that the services offered to employee will not decrease.

To conclude, an employee must genuinely agree to microchipping and have a real possibility to refuse this new technology. We would like to emphasize the fact that the importance of the consent must be highlighted to the employee and they must be meaningfully informed and have a proper knowledge about the technology and monitoring practises used by the employer. The removal of the chip must be done with the same easiness as it was placed. If any of the requirements of consent mentioned prove to be impossible or very difficult to demonstrate, it probably means that there should be a reliance on other grounds to process the data received from chips.

Also, due to the complex and vague rules related to consent in the GDPR, microchips are likely to expand without clear direction, which may further deteriorate employee privacy protections. We believe that, employees and employers would benefit from a clear and stable approach concerning microchipping. Thus, due to the imbalance of power between employers and employees, we argue, that it should be for national laws to strengthen employees' ability to reject microchipping and give specific rules in case of monitoring microchipped employees. Considering that microchipping of employees should be truly optional and voluntary, it might be appropriate to consider adopting a legislation that would prohibit the compulsory use of chips or that requires companies to always provide the infrastructure supporting the use of different technologies and not to rely on chips exclusively.

4 Do Employers Have a Legitimate Interest to Monitor Microchipped Employees?

Concerns around the legal feasibility of consent have lead researchers to consider moving to an alternative legal basis in case of employee monitoring.[57] We propose that, due to its implications, consent should be used together with another lawful basis for data processing in case of microchipping. Consent remains the only mechanism through which the employee can exercise control over their personal data prior to processing. Therefore, we cannot imagine lawful situations where employees are implanted with a microchip despite their refusal. The latter practice would surely infringe employee's bodily integrity and human dignity and be in violation of basic human rights.

As the employer is allowed to possess some degree of authority over the employees due to the nature of the employment relationship, the possible legal

[56] Ibid. Martins Esfola (2018).

[57] Custers and Ursic (2018).

ground that compliments consent can be legitimate interests of the employer.[58] Under this legal ground employers may process employees' data when their commercial interests are so strong that they prevail over employees' rights such as privacy and data protection.[59]

According to Article 29 Working Party, to rely on this legal basis, employer has to carry out a proportionality test and balance their legitimate interest against the interests and fundamental rights of the data subject. The balancing test includes establishing potential consequences for the employee—including the creation of privacy-related risk and an estimation of the likelihood of the materialisation of the risk and its potential severity.[60] The legitimate interests ground, along with the other grounds apart from consent, also require a necessity test,[61] which means that employer needs to show that the use of microchips is necessary and that there are no other means of monitoring least intrusive to the employee. In general, the more negative or uncertain the impact of the processing might be and the more intrusive the monitoring measure is, the more unlikely it is that the processing will be considered, on balance, as legitimate.

Unfortunately, this balancing exercise has been criticized due to its broadness and lack of understanding as the test leaves much room for interpretation.[62] The difficulty with the balancing test is that a balancing of rights is never clear-cut and depends on specifics of the case. As the balancing test is carried out by the employer, it is questionable if the test strikes a fair balance between the parties of the employment relationship and is not biased. The balancing test also needs a strict and substantive analysis that may be complicated for employers to carry out.

Helpful guidance in this matter can be found in the case law of the European Court of Human Rights (ECtHR). In a case *Bărbulescu v Romania*[63] the court conducted a proportionality analysis and shed light on a number of relevant criteria for domestic courts to assess when dealing with employee privacy. This criteria can be summarized as follows: (i) the existence of a prior notification on the possibility to implement monitoring measures and their implementation; (ii) the extent of the monitoring and intrusion into the employee's privacy; (iii) if a legitimate aim was pursued by the employer that could justify the carried monitoring; (iv) whether it was not possible to use a less intrusive method than the one deployed; (v) the ultimate consequences to the employee subject to the monitoring and the subsequent uses the

[58]Custers and Ursic (2018). European Parliament and Council Regulation (EU) 2016/679. Article 6 (1)f.

[59]Under GDPR employers can legitimise employee monitoring, if processing is necessary for the purposes of the legitimate interests pursued by the employer, except where such interests are overridden by the interests or fundamental rights and freedoms of the employee.

[60]Article 29 Data Protection Working Party (2014).

[61]Article 29 Data Protection Working Party (2014).

[62]Article 29 Data Protection Working Party (2014) and Martins Esfola (2018).

[63]Bărbulescu v Romania (2017) The Grand Chamber Judgement of the European Court of Human Rights. No 61496/08.

employer will give to the results of the monitoring system adopted; (vi) and if proper safeguards had been provided to the employee.[64]

Examples of this balancing test can also be found from other parts of the world. The Federal Court in Australia, for example, gave a four part test that can also be used to assess the appropriateness of using chips for employee monitoring purposes. The test asks the following questions: if the measure is demonstrably necessary for meeting a specific need; is it likely to be effective in meeting that need; is the loss of privacy proportional to the benefit gained, and is there a less privacy-invasive way of achieving the same end?[65]

In the light of these court cases, using legitimate interests as a lawful bases of microchipping can be quite complicated as using microchips may lead to very serious consequences for the employee. For example, if the employer starts to gather real-time data from chips and capture exit and entry sequences, location, speed and movement of the employee. This kind of systematic reading of a chips can expose the habits of employees and thus possibly establish their profile. Continuous monitoring of employees constitutes a serious interference with their right of privacy[66] and as a consequence, legitimate interests should not be relied on as a legal ground for the processing.

Also, despite the helpful guidance given by the Article 29 Working Party (now European Data Protection Board, "EDPB") and relevant case law, there is still no readily available explanation to assist employers who plan to microchip their employees. We argue that such explanations are needed as addressing these issues allows employers to determine whether it is appropriate to use microchips in workplace and fosters the development of good practices before privacy invasive ones become entrenched. The following are two examples of situations where microchipping and monitoring chip's data can be considered lawful a under legitimate interest.

In our first example, a company implants employees with microchips to provide them with access to rooms and lockers containing hazardous chemicals. The purpose of the chip is to block unauthorized access to chemicals and monitor their use by the worker to guarantee employees safety and preclude the misuse (e.g. theft, misplacement) of chemical substances. In this case, if microchipping employees has substantial safety benefits, the use of microchips is lawful. Especially in the event of possibly preventing an accident and loss of lives. Here, the impossibility to be able to detach the microchip from the body can be a positive aspect as it prevents employees from taking off the chip and causing dangerous situations.

Moreover, we believe that in case of microchipping the specific nature of the sector and the job position will play an important role in determining whether the rights of employment parties are balanced and the level of intrusiveness is reasonable. As the employer's responsibility is to ensure the safety and health of ones'

---

[64]Bărbulescu v Romania (2017) §121.

[65]Eastmond v. Canadian Pacific Railway (2004) The Judgement of the Federal Court of Australia. No FC852.

[66]Ibid. Article 29 Data Protection Working Party (2017).

professionals and clients, microchipping might be more easily accommodated for certain occupations e.g. locating police officers, soldiers or firefighters who need to be rescued in an emergency, providing secure restricted access to vaults containing highly sensitive data or preventing accidents in laboratories, mining sites or for example psychiatric clinics. In these cases employer may argue that inserting microchips in employees' hand is necessary and the least intrusive means for employee monitoring as alternative solutions are not that effective.

Our second example comes from a company which implants office workers with microchips so as to allow them to unlock doors, access office equipment and pay for snacks from the vending machine. The only purpose of the chips is to give employees easy and convenient use of different services and the employer only accesses data from the chip in case of a complaint initiated by the employee (for example access to a computer was not granted). The employees, who do not want to get a microchip, use an access card with similar functions instead of a chip.

However, in case of the second example, can an employer argue that inserting microchips in employees' hand is necessary and the least intrusive means for employee monitoring as alternative solutions for identification exist (such as access cards or wearable tokens)? There is no simple answer to this question. In our opinion, the processing of data from microchips in this second example may be allowed under legitimate interests as the companies monitoring practise is rather limited, the chip serves the same functions as an access card and employees have the possibility do choose between an access card and a chip.

In both of the examples given above, employers can also take additional actions to mitigate or reduce the scale and impact of the data processing through microchips. Employers have to apply both technical and organisational measures, for example, extensive use of anonymisation techniques, aggregation of data, privacy-enhancing technologies to comply with GDPR. The more significant the impact on the employee, the more attention should be given to relevant safeguards.[67] Most importantly employers can strictly limit on how much data is collected from chips and immediately delete data after use.

If in the examples shown above, the employer has relied on two grounds for data processing—consent and legitimate interests of the employer—we believe the balance of rights might be achieved. Still, for the monitoring to be lawful, the employee has to have genuinely agreed upon the use of this new technology and the interests of both parties have to be balanced, otherwise the microchipping is unlawful. Furthermore, for striking a balance, employer also has to use safeguards and ensure that the collected data are not used for illegitimate further processing (e.g. tracking and evaluation of employees).

Hence, as microchips become more widespread in the workplace, we emphasise the need for specific rules that regulate microchipping employees. It is worth noting that an organization is not allowed to track employee's movements, but thanks to

[67]Ibid. European Parliament and Council Regulation (EU) 2016/679. Article 17; also Article 29 Data Protection Working Party (2014).

microchips such information is potentially available to employers if a need arises in the future. For example, microchips could potentially be paired with location, such as GPS, and other more intrusive technologies to extract more data from the employees. Hence, these new functionalities of microchips may also increase the likelihood of abuse through technology. In these cases, national law should regulate the use of microchips, e.g. to allow microchips to be used only in case of specific occupations and jobs where the need for employer's control is more crucial.

5 Conclusion

Although it is currently difficult to predict how popular microchipping may become as a means to monitor employees' activities, various organizations in different parts of the world have already started to accommodate such practice. Thus, rather than remaining doubtful about the usefulness of such technology in work contexts, researchers and policy makers should start the discussion on how to manage such chips in order to respect the privacy of employees. This chapter aims to start such a discussion by questioning the applicability of GDPR in the context of microchips and the lawfulness of microchipping under the GDPR.

Our analysis indicates that if data received from microchips is linked to an employee who is identified or identifiable, the application of GDPR is triggered. The GDPR will become applicable if the personal data are directly stored on the chip or can be indirectly linked to an employee by combining chip's unique number, with other data.

Due to the physical intrusiveness of chips, it is also important to make sure that employers cannot pressure or otherwise exclude employees who do not want to be microchipped. Such activity would raise fundamental human rights concerns, including bodily integrity. Therefore, employers who want to gather data from microchips have to rely on a valid consent to lawfully introduce microchips in the workplace. It is advisable for the employer to clearly inform employees about the voluntariness of giving their consent.

However, due to several concerns raised in this chapter, we argue that consent should not be considered as the only legal ground for data processing in case on microchipped employees but rather applied together with another legal ground—the legitimate interest of the employer.

The legitimate interest together with the agreement of the employee seems to offer some room for employers who wish to conduct employee' monitoring via data gathered through microchips. It enables the employers to carry out more invasive forms of surveillance as long as they can prove that the measure is not disproportionate and appropriate technical and organizational measures are in place to safeguard the rights of the implanted employees.

An option that complements the general framework provided by the GDPR can be specific national rules that protect microchipped employees. We argue, that due to the complex and vague rules of the GDPR, microchips are likely to expand without clear direction, which may further deteriorate employee privacy protections. Thus we believe that employees and employers would benefit from a clear and stable approach concerning microchipping. In fact, due to the imbalance of power between employers and employees, we argue, that national laws should strengthen the employees' ability to reject microchipping and give specific rules about monitoring microchipped employees. Considering that microchipping of employees should be truly optional and voluntary, it might be appropriate to consider adopting a legislation that would prohibit the compulsory use of chips or require companies to always provide the infrastructure supporting the use of different technologies rather than relying exclusively on chips.

In the future, employee's movements and other sensitive information could be available to employers as microchips could potentially be paired with GPS and other more intrusive technologies to extract more data from the employees. Hence, these new functionalities of microchips may also increase the likelihood of abuse through technology. To prevent this, we argue that national law should regulate the use of microchips, so that microchips could be used only in case of specific occupations and jobs where the need for employer's control is more crucial.

References

Article 29 Data Protection Working Party (2014) Opinion 06/2014 on the notion of legitimate interests of the data controller under Article 7 of Directive 95/46/EC. WP 217. https://ec.europa.eu

Article 29 Data Protection Working Party (2017) Opinion 2/2017 on data processing at work. WP 249. https://ec.europa.eu

Article 29 Working Party (2007) Opinion 4/2007 on the concept of personal data. WP 136. https://ec.europa.eu

Article 29 Working Party (2017) Guidelines on consent under Regulation 2016/679. WP259 rev.01. https://ec.europa.eu

Belli L et al (2017) Selling your soul while negotiating the conditions: from notice and consent to data control by design. Health Technol 7(4):453–467

Custers BH, Ursic H (2018) Worker privacy in a digitalized world under European Law. Comp Labor Law Policy J 39(2):323–344

Custers B et al (2013) Informed consent in social media use: the gap between user expectations and EU personal data protection law. SCRIPTed 10:4 435. http://script-ed.org/?p=1232

Fischer-Hübner S, Hedbom H (eds) (2008) A holistic privacy framework for RFID applications. FIDIS Deliverables D12.3. http://www.fidis.net

Gauttier S (2019) 'I've got you under my skin' – the role of ethical consideration in the (non-) acceptance of insideables in the workplace. Technol Soc. https://doi.org/10.1016/j.techsoc.2018.09.008

Glaser A (2017) The next evolution in office working could be employees getting implanted with a microchip. Vox Media. https://www.vox.com

Graveling R et al (2018) The use of chip implants for workers. European Parliament. http://www.europarl.europa.eu. Accessed 12 July 2019

Halperin D et al (2008) Security and privacy for implantable medical devices. IEEE Pervasive Comput, Special Issue on Implantable Electronics 7(1):30–39

Information Commissioner's Office (2012) What is personal data? – A quick reference guide. https://ico.org.uk. Accessed 12 July 2019
Kanarbik L (2018) Võtmed naha all ehk Miks Ülemiste linnaku töötajad end kiibistada lasevad. In: EestiPäevaleht. https://epl.delfi.ee
Khan NA (2015) RFIDs chip implants and their related ethical issues. https://doi.org/10.13140/RG.2.1.3717.6166
Mangan D (2018) Online speech and the workplace: public right, private regulation. Comp Labor Law Policy J 39(2):357–388
Martins Esfola JP (2018) Bar coded at work. An analysis of the European privacy and data protection legal frameworks regarding the hypothetical use of RFID implants for workers' surveillance. University of Tilburg
Meints M (ed) (2007) D3.7 A structured collection on information and literature on technological and usability aspects of Radio Frequency Identification (RFID). http://www.fidis.net/fileadmin/fidis/deliverables/fidis-wp3-del3.7.literature_RFID.pdf. Accessed 14 Apr 2020
MHI (2019) Automatic Identification and Data Collection. http://www.mhi.org
Mitrou L (2019) Data protection, artificial intelligence and cognitive services: is the General Data Protection Regulation (GDPR) 'Artificial Intelligence-Proof'?. https://www.academia.edu/. Accessed 13 July 2019
Office of the Privacy Commissioner of Canada (2008) Radio Frequency Identification (RFID) in the workplace: recommendations for good practices. https://www.priv.gc.ca/en/about-the-opc/what-we-do/consultations/completed-consultations/rfid/. Accessed 14 Apr 2020
Ogriseg C (2017) GDPR and personal data protection in the employment context. Labour Law Issues 3(2). https://doi.org/10.6092/issn.2421-2695/7573
Padi J (2018) Big data analytics, consent and the European Union (EU) General Data Protection Regulation 2016 (GDPR) - the fallacy of consent and control. Dissertation, Keele University
Poullet Y (2009) Data protection legislation: what is at stake for our society and democracy? Comput Law Secur Rev 25:211–226
Rodney IP et al (2008) The social implications of humancentric chip implants: a scenario - 'Thy chipdom come, thy will be done'. Collaborative Electronic Commerce Technology and Research. IEEE Computer Society, Spain
Rodriguez DA (2019) Chipping in at work: privacy concerns related to the use of body microchip (RFID) implants in the employer-employee context. Iowa Law Rev 104(3):1581–1611
Roosendaal AP, Kosta E (2008) A study on ICT implants. FIDIS Deliverables D12.6. https://research.tilburguniversity.edu
Schmidt G, O'Connor KW (2018) Social media, data privacy, and the internet of people, things, and services in the workplace: a legal and organizational perspective. In: Simmers CA, Anandarajan M (eds) The internet of people, things and services - workplace transformations. Routledge, New York. https://doi.org/10.4324/9781315182407-6
Solove DJ (2013) Privacy self-management and the consent dilemma. Harv Law Rev 126:1880
Special Eurobarometer 359 (2011) Attitudes on data protection and electronic identity in the European Union. http://ec.europa.eu
Suder S, Siibak A (2017) Employers as nightmare readers: an analysis of ethical and legal concerns regarding employer-employee practises on SNS. Baltic J Law Polit 10(2):76–107. https://doi.org/10.1515/bjlp-2017-0013
Thiesse F (2007) RFID, privacy and the perception of risk: a strategic framework. J Strateg Inf Syst 16(2):214–232. https://doi.org/10.1016/j.jsis.2007.05.006
Tikkinen-Piri C (2017) EU General Data Protection Regulation: changes and implications for personal data collecting companies. Comput Law Secur Rev 34(1):134–153. https://doi.org/10.1016/j.clsr.2017.05.015
World Economic Forum (2017) Global shapers survey. http://www.shaperssurvey2017.org. Accessed 12 July 2019

Electronic Personhood: A *Tertium Genus* for Smart Autonomous Surgical Robots?

Tomás Gabriel García-Micó

Contents

The author is a member of a consolidated regional Research Group in Private Law ("Grup de Recerca en Dret Patrimonial", 2017 SGR 1636) funded by the Agency for Management of University and Research Grants, dependent of the Catalan Government (AGAUR—Agència de Gestió d'Ajuts Universitaris i de Recerca), and directed by Prof. Dr. Josep Ferrer Riba (Universitat Pompeu Fabra, Barcelona). PhD researcher and teaching assistant at the Universitat Pompeu Fabra (Barcelona). The author holds a grant for the recruitment of early-stage research staff (FI 2019) awarded by the AGAUR under an open and competitive process and co-funded by the European Social Fund's Operative Program of Catalonia 2014-2020 CCI 2014ES05SFOP007.This article is included within the research program of the Programa Estatal de Investigación, Desarrollo e Innovación orientada a los Retos de la sociedad "Responsabilidad civil y mercado. La compensación del daño económico', whose director is Prof. Dr. Carlos Ignacio Gómez Ligüerre, funded by the Ministry of Science, Innovation and Universities and the FEDER (DER2017-82673-R).
I want to thank Prof. Dr. Sonia Ramos González and Prof. Dr. Lela Mélon (Private Law Area. Universitat Pompeu Fabra) for the time devoted to discussing this issue with me. And to Prof. Dr. Martin Ebers and Prof. Dr. Marta Cantero Gamito for their useful comments during the Doctoral Seminar "Algorithmic Governance and Governance with Algorithms: Ethical and Legal Challenges", that was held at the University of Tartu School of Law on May 23, 2019.

T. G. García-Micó (✉)
Universitat Pompeu Fabra, Barcelona, Spain
e-mail: tomasgabriel.garcia@upf.edu

M. Ebers, M. Cantero Gamito (eds.), *Algorithmic Governance and Governance of Algorithms*, Data Science, Machine Intelligence, and Law 1,
https://doi.org/10.1007/978-3-030-50559-2_5

Abstract Back in 2016, the Committee on Legal Affairs of the European Parliament published a pioneering initiative, the Draft Report with recommendations to the Commission on Civil Law Rules on Robotics (the "Resolution", or the "EP proposed rules"). The EP proposed rules were intended to bring a common EU solution to the legal challenges posed by, amongst others, smart autonomous robots. This chapter scrutinizes the applicability of one of the solutions proposed by the EP proposed rules: the granting of electronic personhood to such robots by critically evaluating the existence of a legal basis for it.

1 Introduction

That technology is evolving faster than legal frameworks do is a well established premise. Recently heightened interactions between artificial intelligence (AI) and the medical science[1] have evolved, and the European Parliament accordingly aspired to create a set of rules dealing with the legal implications of emerging technologies. This chapter deals with the critical appraisal of one of the envisaged solutions proposed by the European Parliament resolution of February 16th, 2017 with recommendations to the Commission on Civil Law Rules on Robotics (2015/2103 (INL)):[2]

> f) creating a specific legal status for robots in the long run, so that at least the most sophisticated autonomous robots could be established as having the status of electronic persons responsible for making good any damage they may cause, and possibly applying electronic personality to cases where robots make autonomous decisions or otherwise interact with third parties independently.

Accordingly, electronic personhood is envisaged as a possible development to be applied to the most sophisticated robots, which are by definition fully autonomous. Applicable to the field of surgical robots, this proposal excludes the currently existent surgical robots that, mostly, pertain to the category of teleoperated robots, master-slave systems[3] or telerobots.[4] All of these nominations depict systems where

[1]See: Houssami et al. (2017) and Ramesh et al. (2004).

[2]Hereinafter referred as to the "Resolution", or the "EP proposed rules".

[3]For the purposes of this chapter, the author will use the classification provided by Wolf et al. (2009), according to which medical robots can be classified as passive, semiactive and active medical robotic systems; as well as teleoperated systems, where "robotic system follows the operator's (surgeon's) hand motions from an offsite control console", Wolf et al. (2009), p. 1397. Other ways of classifying medical robots can be found at Romero Otero et al. (2007), p. 336.

[4]Moustris et al. (2011), p. 376.

the surgeon controls remotely the robot from a console, and the robot obeys the orders of the surgeon. Under the current state of the art, we would have a human being that may be found liable for the harm caused by these robots and electronic personality would no longer be necessary to grant legal relief to the victim. But the current system reveals several other issues, linked with the plaintiff carrying the burden of proof of where the damage occurred (e.g. it was a defect of the robot that caused its malfunction, it was a defect in the code of the robot, it was a negligent act of the surgeon that manipulated the robot from the console, etc.).[5]

Even though the scenario in which such novel solutions may be necessary is, still, a bit precipitated considering the current stage of surgical technology, a critical assessment of novel solutions is not too early given the speed of technological development. The present work disagrees with the EP proposed rules' development: while personhood to some extent eases the liability issues, it also creates additional issues, the legal consequences of which have not been properly considered by these rules.[6] Granting personhood carries with it additional rights and obligations such as the possibility of assigning assets to it,[7] having private property, paying taxes, having constitutional rights and duties,[8] etc.[9] Alternative solutions, mentioned in the proposed rules, are better suited for resolving the liability issue than bestowing autonomous robots with legal personhood, a sort of *tertium genus* between the natural personhood and the legal personhood.

Teubner partially synthesizes the pertinent question of the present work with the following statement: "Private law is therefore faced with a choice: Either it gives autonomous software agents an independent legal status and treats them as responsible actors, or there is an increasing number of "accidents", without anyone being responsible for them".[10] The question can be answered with proposing a third

[5]In fact, according to European Commission (2018), p. 27, the 53% of all the cases rejected by national Courts when dealing with product liability cases (one of the potentially applicable regimes to the cases of harms caused by robots) are related with the burden of proof, whether the plaintiff failing to provide the Court with enough evidence about the defect (32% of the cases, 102) or about the causal link between the defect and the harm caused to the end user (21% of the cases, 67). If this already exists in the area of product liability, this problems may be even worse if we take into consideration the specificities of this concrete area where there are several potential successive causes that may have caused the harm to patient.

[6]To have the complete framework of the debate around personhood on artificial intelligence and autonomous robots, see: Solum (1992), Karnow (1996), Allen and Widdison (1996), Teubner (2006), Chopra and White (2011), Teubner (2018), Pagallo (2018), Ercilla García (2018), Núñez Zorrilla (2019) and Wagner (2019).

[7]See: Palmerini et al. (2014), p. 24; Ebers (2019), p. 21.

[8]This position is shared by Koops et al. (2010), p. 516; Ercilla García (2018), p. 14; Núñez Zorrilla (2019), p. 57.

[9]Some autors, like Solum (1992), pp. 1239–1240, 1243–1253; Gray (1921) use philosophical criteria to determine which entities shall be granted personhood, and they find that these are "intelligence, free will, consciousness, intentionality, and emotions", Wagner (2019), p. 594, something that Wagner describes as not more than a "laundry list", Wagner (2019), p. 594.

[10]See Teubner (2018), p. 109.

alternative: a complete review of the existing legal systems to adjust them—if necessary—to the reality and the challenges that new technologies pose to them. In the author's view, the most appropriate solution is offered by last option.

This chapter is structured as follows: in Sect. 2, the current state of the art on surgical robots is explained to provide up-to-date information on the notions analysed in subsequent sections. In Sect. 3, the EP proposed rules are summarized as one of the most recent EU legal initiatives on the question. Section 4 is devoted to studying the feasibility and necessity of granting electronic personhood on autonomous robots, critically appraising the existing doctrine, as well as providing a potential solution which differs from the EP proposal. Section 5 concludes and provides reflections upon a currently theoretical yet topical debate.

2 The Current State of the Art of Surgical Robots

Surgical robots are a new step in the technological evolution of medicine and surgery. A robot has been in the surgical field defined as a "computer-guided mechanical interface,"[11] which is "reprogrammable" and "equipped with sensors and actuators".[12]

Notwithstanding the latest developments,[13] the current state of the art is represented by *da Vinci* surgical system, a master-slave system[14] that is being actively used in many hospitals. The following table shows the evolution of the installed base of *da Vinci* around the world since September 30, 2018 until September 30, 2019.[15]

As seen in Table 1, the number of *da Vinci* systems installed worldwide has increased for a 12.3% in a year, enlarging the number of installed robots from 4814 to 5406, making the liability questions more than purely theoretical. Moreover, according to the data published at the FDA MAUDE database, 5420 adverse effects have been notified—out of 1,745,000 surgeries performed—related to surgical

[11] See Cadiere et al. (2001), p. 1467.

[12] See Camarillo et al. (2004), p. 2S.

[13] See Aruni et al. (2018), p. 221.

[14] According to the classification of robots provided by Romero Otero et al. (2007), p. 336, that distinguishes between "precise path systems", "intern replacement surgical robots" and "master-slave devices". Other authors use other criteria to classify surgical and medical robots. For example, Wolf and Shoham (2009), distinguish between passive, semiactive and active medical robotic systems; as well as a fourth category of teleoperated systems (where the da Vinci falls), where the "robotic system follows the operator's (surgeon's) hand motions from an offsite control console", Wolf and Shoham (2009), p. 1397.

[15] See Intuitive Surgical (2018), p. 32; Intuitive Surgical (2019a), p. 12; Intuitive Surgical (2019b), p. 11.

Table 1 The *da Vinci*'s installed base since September, 2018 until September, 2019

	September 30, 2018	March 31, 2019	September 30, 2019
United States	3110	3283	3459
Europe	821	893	936
Asia	629	661	718
Rest of the world	254	277	293
Overall	*4814*	*5114*	*5406*

procedures performed with *da Vinci* system in a period between the years 2000 and 2013.[16]

Despite the fact that *da Vinci*, the paradigm of a master-slave device,[17] is the most used surgical robot worldwide, fast developments in the medical field provided new highly developed surgical robots. Smart autonomous robots[18] are machines that are no longer teleoperated, but rather integrated with artificial intelligence (AI), and machine learning (ML) capabilities[19] allowing them to act independently, as they are bestowed with the "ability to take decisions and implement them in the outside world, independently of external control or influence".[20]

Due to the novelty that surgical robots in general represent for the existing legal systems, legal scholarship has yet to deal with the specificities of the concept of autonomy of these kinds of robots. By way of example, in the case of autonomous vehicles possessing similar characteristics, progress has been made in the field of classification or extent of the machine autonomy. The Society of Automobile Engineers identified five levels of autonomy on autonomous vehicles: from a no-automation (or level 0) as the level in which "the system automatically assists the driver to regain lost control of the vehicle",[21] the level 1 or function-specific automation, when the 'system controls one function",[22] the level 2 or combined

[16]See Alemzadeh et al. (2016), p. 9. Of these 5,420 adverse effects 126 (2.32%) did end with the death of the patient and 1224 (22.58%) caused non-deadly injuries to the patient.

[17]See Romero Otero et al. (2007), p. 336.

[18]Or "intelligent robot", under § 2.28 of the ISO rules number 8373:2012, International Organization for Standardization (2012). We can see how the feature of machine learning appears on the definition of an intelligent robot, by making them learn from external stimuli to apply their learning to the reality, and not only to learn, but also to adapt their own behaviour.

[19]According to Pagallo (2018), p. 3, even though we are not in a position that robots substitute humans in performing their jobs or tasks, "we are dealing with machine-learning systems", which affects ther decision-making processes, their capability of learning from and respond to external stimuli.

[20]Delvaux (2017), pp. 6–7. In a similar way, other authors define intelligent robots as machines with the ability to "perceive something complex and make appropriate decisions", Singer (2009), p. 77. See also Palmerini et al. (2014), pp. 15–17, where the three different definitions of "robot" are summarized. See also, in general about the concept of artificial intelligence, Muñoz Villarreal and Gallego Corchero (2019), pp. 69–71; Rusell and Norving (2008), p. 2.

[21]See Karanasiou and Pinotsis (2017), p. 5.

[22]Karanasiou and Pinotsis (2017), p. 5.

function automation, when the 'system controls at least two functions",[23] level 3 or limited self-driving automation, when "the driver cedes full control under specific conditions",[24] and level 4 or full self-driving automation, when the "driver is not expected to become involved throughout the duration of the trip".[25] Scientific scholarship on the other hand shies away from such classification,[26] arguing that "this area of research is very dynamic and, in our opinion, has not yet reached maturity, it is likely that categories and classifications that are widespread today will change over time, with the evolution of new concepts".[27]

The scientific scholarship limits the classification to four concepts of robots depending on its autonomy (and disregarding a fifth one called navigators): passive, semiactive, active and robotic manipulators (or "telerobotics"[28]). Passive robots have been defined as systems that "support the surgical procedure, but take no active part during surgery",[29] semiactive robots as systems that "provide guidance of the surgical tool but the actual surgical operation is conducted by the surgeons",[30] while active robots "perform surgical tasks, such as drilling or milling, autonomously with no direct intervention of the surgeon."[31] The notion of remote manipulators encompasses machines that follow "the operator's (i.e. the surgeon's) hand motions from an offsite control console that can be located either in the operating room or even somewhere overseas using fast communication lines."[32]

As technologies are continuously evolving at a faster rate than the scholarship and legal frameworks are, we are entering a new age of surgical robotics with examples such as the STAR® (or Smart Tissue Autonomous Robot), defined as semi-autonomous,[33] but that "[a]fter registering the markers, the robot autonomously performs the tumor cutting".[34]

The abovementioned scenario calls for solutions to both current legal problems that may arise while using the *da Vinci* or any other teleoperated surgical robot, and those that may arise while using cutting-edge technology robots bestowed with AI and ML capabilities, like the project envisaged by Verb Surgical.[35]

---

[23] Karanasiou and Pinotsis (2017), p. 5.

[24] Karanasiou and Pinotsis (2017), p. 5.

[25] Karanasiou and Pinotsis (2017), p. 5.

[26] See Moustris et al. (2011).

[27] See Wolf et al. (2009), p. 1398.

[28] See Moustris et al. (2011), p. 376.

[29] See Wolf et al. (2009), p. 1398.

[30] Wolf et al. (2009), p. 1399.

[31] Wolf et al. (2009), p. 1399.

[32] See Wolf et al. (2009), p. 1400.

[33] See Opfermann et al. (2017).

[34] See Strickland (2017).

[35] This is what the scholarship calls as the problem of "regulatory connection", Brownsword and Goodwin (2012), pp. 63 et seq., pp. 371 et seq. This situation can be presented in two different scenarios: (1) if appears a completely new technology, or if (2) an already marketed technology

3 The Origins of the Debate: The EP Proposed Rules

The purpose of the EP proposed rules is to harmonize the EU Civil Law applicable to robotics. It does not entail a harmonization of EU Tort Law, but a sectorial substantive change in several areas of the law, addressing issues of civil law liability, private international law, regulatory law, health law, employment, and education law, environmental law, defense law, and fundamental rights and ethics.

On May 31st, 2016, Mady Delvaux, member of the Party of European Socialists and the JURI Committee (the Committee on Legal Affairs) filed, as the rapporteur, a draft report with the EP proposed rules, containing several material proposals. Firstly, the proposal aimed at solving the legal issues arising from the use of artificial

mutates or changes. In both cases, it is considered that regulation hardly stays connected to the technological reality and that that should be dealt with a process of envisaging a regulatory framework, but that such regulation needs to balance two different interests: legal certainty (for patients, producers, doctors, etc.) and not to discourage innovation and investments in research of new technologies due to an excessively burdensome legal framework, Palmerini et al. (2014), pp. 10–11. It should be also mentioned that the lack of regulation is the premise over which the EU Parliament's JURI Committee builds the Report: at the final report on the EU Civil Law Rules in Robotics, we can see how the rapporteur (Mady Delvaux) did pinpoint the feasibility of a scenario of autonomous robots, and that scenario has the problem that "the traditional rules will not suffice to give rise to legal liability for damage caused by a robot, since they would not make it possible to identify the party responsible for providing compensation and to require that party to make good the damage it has caused", Delvaux (2017), p. 7. Also the report issued by the Directorate-General for Internal Policies introduced, as a preliminary remark, that "civil liability law, for example, might be less easily applied to developments in autonomous robotics, particularly in a scenario where a machine might cause damage that cannot be easily traced back to human error. Whole chapters on civil liability might, then, need rethinking, including basic civil liability law, accountability for damage, or its social relevance", Directorate-General for Internal Policies (2016), p. 6. Quite afterwards, we can see how the Directorate-General highlights six questions that may be the grounds of an interesting debate on civil liability law, such as the introduction of a "vicarious liability [regime] for the robot(s)", Directorate-General for Internal Policies (2016), p. 16, or the difficulty of determining liability for damages caused by an autonomous robot, because "it may be more difficult to ascertain what caused the damage in certain situations, particularly if the robot is able to learn new things by itself", Directorate-General for Internal Policies (2016), p. 16, or the need of not splitting up the civil liability regimes applicable to autonomous and non-autonomous robots, because that will pose on judges a technical duty of checking "whether the robot corresponds to the definition of a smart robot in the present motion and to determine the applicable law", which will, in turn, suppose that the parties will hire "an expert to determine the machine's capabilities. This would lead to a considerable increase in procedural costs and slow down legal proceedings", Directorate-General for Internal Policies (2016), p. 17, amongst other solutions. The analysis of each one of these solutions exceeds the object of this chapter, focused on the analysis of whether one of the solutions that the Report drafted by Mady Delvaux is feasible or not: granting legal personhood to autonomous robots.

intelligence[36] (not specifying whether referring to strong or weak AI[37]) and machines equipped with machine learning capabilities.[38] Secondly, the pertinent rules on liability have been qualified as "insufficient,"[39] a situation that "calls for new rules which focus on how a machine can be held – partly or entirely – responsible for its acts or omissions."[40] Thirdly, and relating to the discussion in Sect. 2 above, the proposal calls for a discussion on "whether [the intelligent robots] should be regarded as natural persons, legal persons, animals or objects – or whether a new category should be created, with its own specific features and implications as regards the attribution of rights and duties, including liability for damages".[41] Fourthly, several solutions have been proposed for resolving the issue of civil liability suggesting the use of a strict liability-based regime: (1) a "compulsory insurance scheme", (2) a "compensation fund", similar to already existing ones for compensating non-medical[42] and medical issues,[43] and "allowing the manufacturer, the programmer, the owner or the user to benefit from limited liability insofar as smart autonomous robots would be endowed with a compensation fund", as well as (3) the one that will be the object of this chapter, the creation of a 'specific legal status for robots, so that at least the most sophisticated autonomous robots could be

[36]Regarding the concept of artificial intelligence (AI) there is not a 'straightforward, consensus definition", Calo (2017), p. 4. For a better understanding of this debate, see Rusell and Norving (2008), p. 2; The Headquarters for Japan's Economic Revitalization (2015), p. 21; Calo (2017), p. 4; Puaschunder (2018), p. 2.

[37]Not every AI is equal to each other, but there is a general distinction between a weak AI and strong AI, Searle (1980), p. 417; O'Regan (2016), p. 257; Wisskirchen et al. (2017), p. 3; Puaschunder (2018), p. 3.

[38]See Delvaux (2016), p. 5. Machine learning (ML) is considered as a subfield within the study of artificial intelligence; Surden (2014), pp. 89–90; Calo (2017), p. 4; Muñoz Villarreal and Gallego Corchero (2019), p. 69.

[39]See Delvaux (2016), para. S (p. 5), V, Y-Z (p. 6). A position shared also by the report of the Directorate-General for Internal Policies (2016), p. 6, pp. 16–18, as well as by the scholarship, see Lacruz Mantecón (2019), p. 5. See also Holder et al. (2016a), pp. 7–8.

[40]See Delvaux (2016), p. 5. The issue of robot liability has been also stated by the Spanish Government during the meetings to amend the *Pacto de Toledo* (Toledo's Deal) on retirement pensions. The preliminary deal between the Socialist Party (PSOE) and *Unidos Podemos* (United We Can), which was not finally approved due to the breach of negotiations between both parties, had as one of main points the fact that robots were considered as apt persons to pay taxes, Jorrín (2019).

[41]See Delvaux (2016), p. 5.

[42]In Germany was passed a bill to compensate those who worked in concentration camps during the Nazi regime (*Gesetz zur Zahlbarmachung von Renten aus Beschäftigungen in einem Ghetto* or ZRBG), or the compensation funds for the victims of asbestos in France and Belgium, the Workmen's Compensation Acts" funds and the basic protection systems for car accidents of the United States, or the compensation fund envisaged by the Spanish Law 26/2007, on October 23rd, on Environmental Liability (*Ley 26/2007, de 23 de octubre, de Responsabilidad Medioambiental*).

[43]The *Patienterstatningen* from Denmark, or the several medical compensation funds of the United States existing in Georgia, Indiana, Kansas, Louisiana, Pennsylvania, Nebraska, New Mexico, New York, South Carolina and Wisconsin.

established as having the status of electronic persons with specific rights and obligations."[44]

The Directorate-General for Internal Policies (hereinafter, the "DG IPOL"), issued its report on the EP proposed rules in October 2016.[45] The report analysed the entire motion for a resolution, rejecting the idea of electronic personhood as "unhelpful" and "inappropriate", due to the fact that "[l]egal personality is therefore not linked to any regard for the robot's inner being or feelings, avoiding the questionable assumption that the robot is a conscious being."[46] This statement raises a question if the existing legal persons (i.e. companies, or associations) have "inner being or feelings" or are "conscious"?

The answer to that question by DG IPOL has not been unambiguous: it simply stated that while the law has already assigned legal personality to a non-human entity, such assignation should not be extended to intelligent robots,[47] supporting the argument with the agency theory considerations. It justified the difference between legal persons and robots as follows: "Legal persons are able to act within the legal sphere solely because there is a human being behind the scenes to represent it," and in the case of robots, "[w]e have two options: either a physical person is the true legal actor behind the robot, or the robot itself is a legal actor."[48] In the first case, the "person would represent the electronic person, which legally speaking, would – like the legal person – simply be a fictional intellectual construct,"[49] but the DG IPOL report considers that the motion for a resolution is considering fully-autonomous robots, without human presence. According to the report, such case will be easily solved by using one of the other proposed mechanisms, rather than bestowing personhood on a "mere machine, a carcass devoid of consciousness, feelings, thought or its own will."[50] The final report issued by the JURI committee disregarded these objections and maintained the idea of creating a specific legal status for autonomous robots.[51]

The idea of the creation of an electronic-person status for robots was one of the main points of discussion during the debate held in the European Parliament on February 15th 2017. Some of the most interesting positions were the ones defended by Dita Charanzová, the rapporteur for the opinion of the Committee on the Internal

[44]See Delvaux (2016), pp. 11–12.

[45]See Directorate-General for Internal Policies (2016).

[46]See Directorate-General for Internal Policies (2016), p. 14.

[47]Directorate-General for Internal Policies (2016), p. 14.

[48]See Directorate-General for Internal Policies (2016), p. 15.

[49]Directorate-General for Internal Policies (2016), p. 15.

[50]Directorate-General for Internal Policies (2016), p. 15.

[51]See Delvaux (2017), p. 18.

Market and Consumer Protection on the Resolution;[52] Gilles Lebreton,[53] Pavel Svoboda,[54] Arne Gericke,[55] and João Pimenta Lopes.[56]

The European Parliament approved on February 16th, 2017, its motion for a resolution. It expressly recommended, as a solution to the legal issues posed by the introduction of new technologies with self-learning abilities, and with enough autonomy as to avoid the interaction with a human being, the creation of specific legal status for robots: the electronic person. The following section assesses the feasibility of such a solution.

4 Should Autonomous Robots Be Granted Personhood?

4.1 Preliminary Considerations

As argued above, the granting of personhood to robots is one of the most controversial points of the EP proposed rules. One could say that the EU legislator intends to envisage solutions "in the long run"[57] as fully-autonomous robots do not exist yet, acting in a proactive manner. Given the current trajectory of technological developments, litigation related to the use of teleoperated systems can become a reality in an instant so it is crucial for legal systems (at least within the EU) to envisage solutions ex ante to be able to accommodate such litigation.

In Spain there are currently no cases of litigation for harms suffered while undergoing surgery with the *da Vinci*. The existing litigation rather deals with questions of private health insurance coverage of the expenses related to the surgical procedure itself.[58] Notwithstanding the latter, there exist some litigation on to the

[52]We must reject that we are creating "electronic persons": I am sorry, but robots are not humans, it is pure science fiction to think otherwise.

[53]I protest against this perspective for two reasons: first, because it would exempt manufacturers and users from liability, [...]; next, and above all, because I believe [...] that the human person possesses a spiritual over-existence to which no artificial intelligence can be compared.

[54]I will support the amendments on some points, particularly as regards the creation of a specific legal status for robots or a basic income. I think we would be too far away from predictable reality.

[55]This is nothing else than his right to privacy for robots. Therefore, ladies and gentlemen, please support our amendments four to eight.

[56]At the same time, it takes a very narrow approach to impacts in the world of work, notably increasing precariousness, impoverishment, unemployment and the liquidation of labor and social rights, omitting the interests of capital in its development, to increase exploitation with maximization of profit.

[57]See Delvaux (2017), p. 18.

[58]Judgments of the Court of Appeals of Barcelona, Section 17th, of 24.10.2018 (ECLI:ES: APB:2018:10484; Rapporteur: Mireia Borguño Ventura); of Cáceres, Section 1st, of 17.9.2018 (ECLI: ES:APCC:2018:659; Rapporteur: Juan Francisco Bote Saavedra); Zaragoza, Section 1st, of 6.10.2017 (ECLI:ES:APZA:2017:387; Rapporteur: Ana Descalzo Pino); Murcia, Section 1st, of 27.2.2017 (ECLI:ES:APMU:2017:563; Rapporteur: María del Pilar Alonso Saura); Valencia,

hazardous consequences of the use of a surgical robot in other jurisdictions.[59] The best-known cases in the field are the US cases *O'Brien v. Intuitive Surgical* (Docket: 10 C 3005, District Court for the Northern District of Illinois (East Division)), and *Taylor v. Intuitive Surgical* (Docket: 92210-1, Supreme Court of Washington).[60]

Surgical robots do not only pose a risk to humankind when fully autonomous but also when their use presumes human guidance as traditional legal solutions might not be sufficient or conversely they might lead to an expensive and long litigation procedure for the victims to get compensation. Issues not covered by existing legislation range from issues such as if the intervention of the robot affects or not to the causal link between the harm and the action or omission; or if the robot has the ability to learn and one day during a surgery causes harm to the patient due to a problem not related to the initial code provided to it, who is responsible for these damages? The manufacturer, the coder, the hospital who requested the robot? Additional issues arise as a consequence of the definition of "product" under the Council Directive 85/374/EEC of July 25 1984 on the approximation of the laws, regulations and administrative provisions of the Member States concerning liability for defective products: is a robot a product? Or the traditional scope of this concept may need to be revisited to adapt it to the technological revolution that such systems represent? Answering all of these questions in detail surpasses the scope of this chapter.

Regardless of the considerations presented above, European tort law systems are suited for solving a vast majority of issues arising between private entities, even when those are more complex. The more specific area of product liability law is focused on a concrete image of what a product is, causing issues when more complex products are in questions (not in the legal sense), such as robots, software[61] or blood, etc. In that regard, does the assumption of the EP proposed rules stand on the necessity of these solutions in the long run? The present work argues to the contrary. Solutions are needed now more than ever,[62] as already the current reality demands novel solutions, calling for more legal certainty. Envisaging solutions for a yet-far-

Section 11th, of 30.9.2013 (ECLI:ES:APV:2013:4348; Rapporteur: José Alfonso Arolas Romero); and Gijón, Section 7th, of 26.10.2010 (ECLI:ES:APO:2010:239; Rapporteur: Ramon Ibáñez de Aldecoa Lorente).

[59]For instance, it is worth citing Nik-Ahd et al. (2019). In this chapter, the authors use the Bloomberg Law database to analyse how many cases involving the *da Vinci* were indexed. According to the empirical data provided, 123 claims have been filed since 2000 (when the robotic-assisted surgery was approved by the FDA) until 2017 (14 of them were class action suits), of which 43 (a 34.96%) have been dismissed.

[60]But, as Wagner (2019), p. 604 states, these cases deal with "peripheral questions": the learned intermediary doctrine, the statute of limitations, etc.

[61]For a better analysis on this question, see Stapleton (1989).

[62]A similar position is held by Gunther Teubner on his most recent research. He says that "autonomous software agents are generating previously unknown risks, so that new rules in private law are probably needed to counteract them", Teubner (2018), p. 108.

to-reach future is just a hypothesis, while we are currently already facing situations that require clear rules regarding the already existing technologies (*da Vinci* robot is an example), which should be the legislator's priority.

4.2 *Electronic Personality as a Liability Facilitator?*

Granting legal personhood to robots needs detailed clarification as it is currently unclear whether this personhood amounts to making robots legal subjects with rights and obligations, or conversely serving just the imposition of liability for harms caused to third parties. According to the Expert Group on Liability for New Technologies, it would be possible to create a specific status for these fully autonomous technologies just for the purposes of imposing them liability (without any other kind of right).[63]

Some authors have proposed that such personhood should be limited, equating the robots to the status of slaves at the Ancient Roman Empire,[64] or treating them as agents (under the rules of the Spanish mandate or the Common Law of Agency)[65] of their *dominus*,[66] whose liability will be limited to the amount granted to the e-person as a *peculium*, even by extending the legal nature of the *peculium* far beyond the scope of the contractual liability, where it was applicable according to Roman Law.[67] In the light of latter, two potential solutions have been envisaged by some scholars, discussed herunder.

4.2.1 The Agency Solution: The Robot (E-Person) as a Representative of Its Owner

Under the U.S. law, the application of the rules of agency arguably represents a forceful attempt to secure a solution, as in general these rules require that the agent-to-be and the principal-to-be both express their consent to enter into this fiduciary

[63]See Expert Group on Liability for New Technologies (2019), p. 38.

[64]See Ercilla García (2018), p. 21.

[65]Eventhough there is not a specific mention to which legal regime should be applicable, the autor will use an example of two completely different legal regimes: Common Law (with the U.S. as an example) and the Civil Law (with Spain as an example).

[66]Ibid., 23.

[67]Ercilla García (2018), p. 25, as well as Pagallo, considers that the specificity that robots pose for legal frameworks can be solved by applying, even in tort liability, the regime of the *peculium* when, as it is considered by the scholarship following the *Gaii Institutionum* § 74 and the Justininan *Institutione*, the *actio de peculio* was only applicable to contractual liability and when the *servus* (the slave and the son) entered into a contract following the instructions of the *dominus* of the slave or the *paterfamilias*. See Cassarino (2016), pp. 248–249; Stolfi (2001), pp. 414 et seq.

relationship,[68] except if by force of a statute, the legislator creates the legal fiction of a "forced" agency relationship,[69] similar to the case in which the court appoints a lawyer for a person appearing before the court.

Spanish law provides for a similar institute: the mandate, regulated by articles 1709 ff. of the Spanish Civil Code, constituting a voluntary fiduciary relationship that requires consensus of the parties.[70] Only in exceptional cases the law expressly creates a non-consentual mandate, such as the case of parental authority minors who are not emancipated,[71] guardianship of minors or incapacitated people,[72] defenders in cases of conflicts between guardians and children[73] or absent people.[74] In all of these cases, it is expressly stated that the agent's representation powers are limited to the functions included in the fiduciary relationship.[75] If we consider the agency theories applicable, it needs to be defined what is the 'scope of the agent's actual authority" (under the terms of §8.09 of the Restatement) or the "limits of the mandate" (under the terms of article 1714 of the Spanish Civil Code). One might consider that the agent's (the robot's) scope of authority is the code provided by the programmer, and while a harm has been caused by the robot's compliance with the provided code, the responsibility will be of the principal (the programmer).

[68]See the Restatement of the Law Third: Agency, Comment *c* to §1.01. See, also, B & G Enters., Inc. v. United States, 220 F.3d 1318, 1323 (Fed.Cir.2000) and Judah v. Reiner, 744 A.2d 1037, 1040-1041 (D.C.2000).

[69]Comment *d* to §1.01 of the Restatement of the Law Third: Agency provides examples of this kind 'situations that resemble agency in form but in which the parties" consent is subject to constraints imposed by law or by legal or regulatory institutions", American Law Institute (2005), p. 22: the owner of a lost item (principal) and governmental officials who recover it (agent), and the case of the court-appointed lawyer, being the principal the court and the agent, the lawyer (in this last example, the opinion of the client is irrelevant).

[70]See Bercovitz Rodríguez-Cano (2001), p. 1940.

[71]See article 154.I of the Spanish Civil Code: "Non-emancipated children shall be under the parents' parental authority".

[72]See article 222 of the Spanish Civil Code: "The following persons shall be subject to guardianship: 1. Non-emancipated minors not subject to parental authority. 2. Incapacitated persons when the judgement has ruled it. 3. Persons subject to extended parental authority, upon termination thereof, save in the event that conservatorship should apply. 4. Minors who are in situation of neglect".

[73]In this cases, the Spanish Civil Code expressly excepts the representation powers of the guardians (see article 162.II.2 of the Spanish Civil Code), and such conflict of interests shall be resolved by the children appointing a defender to take care of his or her interests in that precise conflict, both in- and out-of-court (see article 163.I of the Spanish Civil Code).

[74]See article 181.I of the Spanish Civil Code: "In any event, upon the disappearance of the person from his domicile or from his last place of residence, without having any further news on him, the Judge may, at the request of the interested party or of the Public Prosecutor, appoint a defender to protect and represent the disappeared person in court or in any business which does not admit delay without serious detriment".

[75]See §8.09 of the Restatement of the Law Third: Agency and article 1714 of the Spanish Civil Code.

As artificial intelligence features have proven to be difficult to control due to their dynamically adaptive nature,[76] their machine learning abilities might cause the robot deviating from the initial code. If harm is caused by a smart autonomous surgical robot that has deviated from the code granted to it, can the principal (the programmer) be held liable? Following the rules of agency and mandate, the answer would have to be negative, so we are left in a situation where no one is liable for the harm caused. The initial question of the responsibility for a smart robot's autonomous conduct thereby remains unresolved.[77] A second possible solution therefore could be the granting of personhood of robots, consisting of creating a *tertium genus* between the natural person and the legal person, as envisaged in the EP proposal.

4.2.2 Electronic Personhood

Granting personhood to a robot could resolve the legal void arising from the application of the agency rules on an autonomous robot which deviated from its source code, making the robot itself liable.[78]

A separate legal personality of smart autonomous surgical robots would entail a similar limitation of liability as the one granted to companies or associations: the person behind the robot would be exempted of liability, what is commonly called amongst scholars "owner shielding".[79] To ensure that such shielding is practicable, it is necessary that the "person" owns assets[80] and can be held liable and pay the liability that corresponds from any harm done to a third party.

According to some scholars, granting personhood on autonomous robots would ease the burden of proof of one of the hardest issues in cases of new technologies and tort law: the identification of the liable subject.[81] But such a statement is only

---

[76] See Pagallo (2017), p. 20.

[77] These are the "hard cases" mentioned by the scholarship. See O'Sullivan et al. (2019), p. 6. In these cases, Ugo Pagallo considers that a potential solution would be to bring a Roman law institution: revisit the *peculium*. See the idea of the "digital peculium" in Pagallo (2013), pp. 132–134.

[78] According to Ercilla García (2018), p. 20; Núñez Zorrilla (2019), p. 63; Leroux (2012), pp. 60 et seq.

[79] See Hansmann et al. (2006), p. 3. Some authors consider that owner shielding will result in negative consequences, because it reduces the incentives for producers, programmers and every user of the robot of acting with due care, because no liability would be claimed to them, Wagner (2019), p. 609. This statement does not take into account the reputational costs: the users of a defective product will punish producers when a product is not safe enough, Ganuza et al. (2016), pp. 233–235.

[80] See Alfaro Águila-Real (2016), Chopra and White (2011) and Ebers (2019). Specifically this is one of the reasons why the Expert Group on Liability for New Technologies discards the possibility of granting personhood to fully autonomous technologies, Expert Group on Liability for New Technologies (2019), p. 38.

[81] See Holder et al. (2016a), p. 13; Holder et al. (2016b), pp. 6–7; Wagner (2019), pp. 611–612.

partially true: the defect might be in the original code, so the liability can be directly claimed to the programmer; or in the training of the robot, etc.[82]

Just by granting legal personhood to robots, the plaintiff would not be automatically relieved of the burden of proof as to the causation. It would remain to be determined which was the robot's act or omission that harmed the plaintiff, and the robot's attorney will have the possibility of proving that there is a more adequate cause for that damage rather than the robot's conduct. At the end of the day, the issue of the burden of proof would remain. Factually robots' legal personality simply shifts the issue from the identification of the liable subject to causation, which currently also represents an issue[83] in product liability claims.

Aside from the abovementioned issues, there are many other reasons, aside from the philosophical questions,[84] that are legal in nature, furthering the arguments against the legal personhood of robots. Firstly, a fully-autonomous robot would not be endowed with a legal personality for the same reasons than a company is. A surgical robot—without appraising its autonomy—is nothing more than an instrument to perform a surgery on a patient, a tool, while corporate legal personality represented a necessary tool to foster and incentive investing strategies, to strengthen economies and to secure the creation of jobs.

Companies are created to perform an economic activity, to be an actor in the market and to interact with users.[85] Companies exist to perform different activities (such as to offer legal advice, to build, to provide with medical equipment, to transport goods, to transport patients, etc.) and these activities, when performed regularly and according to the existing legal frameworks, may create significant debts due to a poor management of the economic resources of the company, qualified as a simple mismanagement, without an intention to create a company to elude liability. A person who creates a company which goes bankrupted due to mismanagement of its resources cannot be made personally liable[86] for such debts—unless he acted intentionally or created the company only to perform illegal activities or for money laundering, etc. If a government made the person behind a company entirely liable, notwithstanding the cause of the bankruptcy, nobody would be interested in creating a company; jobs will not be created, there will not be any economic activity and countries will be impoverished.

It should also be noted that behind the legal person there is always a natural person, a person who can be made liable as an exception to the rule.[87] In the case of fully autonomous robots, such a link seems to disappear, as the producer and the programmer can be different persons, and they cannot be held liable for actions or omissions of a robot that were not foreseeable at the time of its manufacturing and

---

[82] See Expert Group on Liability for New Technologies (2019), p. 38.

[83] See European Commission (2018), p. 27.

[84] See Chopra and White (2011).

[85] See Alfaro Águila-Real (2016).

[86] See Wagner (2019), p. 598.

[87] See Muller (2017), par. 3.33; O'Sullivan et al. (2019), p. 7.

coding (e.g., when the machine learning abilities of the robot make it to act beyond of the original code.)[88] Beyond a moral issue this is an issue of causation, as one of the issues of the traditional theories of Tort Law.

In the case of surgical robots, an additional question arises: if "machines [...] things, notwithstanding how singular they are; inanimate objects that, despite they can do tasks, as people do, differ with them in several and important aspects"[89] shall be granted with personhood. Some authors consider that such a possibility will "degrade human dignity",[90] allowing for moral, not solely legal, considerations to enter the debate.

When dealing with this debate on the granting of personhood to robots, a careful balancing of legal and moral issues is needed, creating a new legal framework to that effect, called Robot Law.[91]

4.3 On a Possible Alternative Solution

Alternatives to granting personhood on robots could present better outcomes for the problem at hand. It is possible to modify legal frameworks in a way that they provide alternative solutions, determining the person or company that should be held liable for the damage caused by a surgical robot while performing an operation, even when the surgical robot is entirely autonomous; or by finding alternatives to ease the way of the plaintiff to get compensation. Liability, at the end of the day, does not require personhood, but a full adaptation of national and EU legal frameworks to new technologies (an ongoing initiative at the EU level).[92]

---

[88]See Griffin (2017), that explains that Facebook had to shut down two chatbots after they developed their own language, a language that only the chatbots understood and that was incomprehensible for human beings. Let's imagine that an autonomous surgical robot considers that the most efficient way of performing a specific surgery is to cutting through an area that harms another organ. If the code did not contemplate that specific scenario, the coder would not be liable; nor the producer, because it is not a defect on the robot itself.

[89]See Rogel Vide (2017), 2018) and Lacruz Mantecón (2019).

[90]See Yampolskiy (2018). By way of example, Saudi Arabia, the social robot Sophia was granted legal personhood—by granting it citizenship (given that she does not have, in itself, personhood, we know that citizenship is intertwined with personhood. A non-personal entity cannot be bestowed citizenship)—and, according to Yampolskiy, "human women, including feminist scholars, objected, noting that the robot was given more rights than many Saudi women have", Yampolskiy (2018). Again, the distortion between accountability and personhood arises. It is clear that the government of Saudi Arabia, when granting legal personhood on robot Sophia, was not thinking about liability or accountability, but to get the attention of the media—and succeeded as almost every newspaper around the world dealt with the issue—.

[91]Calo (2011, 2017), Richards et al. (2016) and Lacruz Mantecón (2019).

[92]The EU is gathering experts to evaluate the Directive 85/374/EEC (see the Commission Staff Working Document, SWD(2018) 157 final), or to analyze whether the Directive itself is fit for emerging new technologies (see the Commission Staff Working Document, SWD(2018) 137 final).

One potential solution might be the creation of a specific insurance scheme, that turns into a way of revisiting institutions of contract law to assess how the risks and costs posed by emerging technologies can be internalized.[93] This would involve not a general civil liability insurance, but a case-specific insurance,[94] similar in nature to the homeowners insurance, which not only covers the damages that one can suffer in its own dwelling as a result, for example, of a fire; but also the damage that can be caused to other's property due to, for example, the break of a pipe, or due to a fire caused due to the bad maintenance of an electrical household appliance.[95] In such cases, the solution will consist on making manufacturers of smart autonomous surgical robots responsible for introducing unavoidably dangerous products into the market, obliged to obtain a specific insurance to cover any damage that the robot may cause to a third party (the patient) while undergoing a surgery and that it is caused by a defect present on the robot (i.e., manufacturing, design or information defects of any kind).

In the case the harm suffered by the patient was caused by how the robot is operated by the surgeon in charge of the procedure, and not by a defect on the robot (or this is not proved), then the manufacturer will not be held liable, while the plaintiff would have to prove whether the surgeon has been diligent or negligent, or whether a normal or an abnormal functioning of the public health system has taken place.

Example of case-specific insurance schemes can be found in §109 of the California Insurance Code, as amended in 2018, that provides the definition of the "workmen's compensation insurance", that covers "loss from liability imposed by law upon employers to compensate employees and their dependents for injury sustained by the employees arising out of and in the course of the employment, irrespective of negligence or of the fault of either party".

On the other hand, authors like Bertolini consider that when risk management is intertwined with new emerging technologies, the internalization mechanism inherent to insurance contracts does not work as expected as there is a lack of sufficient data to foresee the party involved in the accident, and the nature and severity of the risk to be prevented or compensated. This, in turn, will result in three potential scenarios, which are that insurance companies will exclude certain kinds of robots from coverage, or that they will use the same contracts that are currently used, which will prove to be inadequate; or impose high premiums, hampering technological innovation and investment.[96]

The Expert Group on Liability for New Technologies deals also with this potential solution in its report, adopting a cautious standpoint because they consider

[93]In fact, this is a position shared by Bertolini (2016), p. 296. A position that won't hamper investment in new emerging technologies. According to Bertolini (2016), p. 299, there is not a liability gap, but an overlapping of different legal regimes that lead to uncertainty.

[94]Or, as Bertolini (2016), p. 295, says, a 'third party insurance'.

[95]A similar idea is proposed by Holder et al. (2016b), p. 14.

[96]See Bertolini (2016), pp. 310–312; Bertolini et al. (2016), pp. 387–388.

that this scheme should "not be introduced without a careful analysis of whether it is really needed",[97] because this could bring negative consequences for the development and investment by companies in these new technologies. Notwithstanding, the Expert Group considers that this solution is better suited for some economic sectors, such as transportation, industries with a high potential for personal injury and/or environmental harm, hazardous activities and certain professional sectors.[98]

The health sector where this systems are used (taking into consideration that they are not fully autonomous, yet) could be one of these where there is a high potential for personal injury, not in probability terms, but in the severity of the damages that a patient may suffer as a consequence of a misuse or a defect in the surgical robot used to perform the surgery upon the patient.

It is true that this solution does not ease the hard burden of proof issues related to the use of robots in surgery (different from the personhood solution), but this solution seems to have *prima facie* a stronger legal foundation than creating an electronic-person status for autonomous robots but, as said, is not exempt of other practical issues which surpass the scope of this chapter.

5 Concluding Remarks

The present chapter has dealt with a pertinent issue of the lack of adaptation of national and EU legal frameworks to the new developments of surgical robots use across the EU.

The EP proposed rules are a good initial proposal for finding legal solutions to the issues posed by a rapid evolvement of new technologies. The debate among the scholars highlights the weaknesses of rationales used to defend the idea of granting legal personality to autonomous robots, at least in the surgical field, when solutions to the current state-of-the-art technologies are needed.

The present work argued that granting legal personality to robots is unnecessary. Robots do not need to be held responsible, neither do they need to be granted personhood. A third option has been suggested, one of adapting the existing legal frameworks to the challenges posed by the advanced new technologies, such as surgical autonomous robots. While the need for granting legal personhood to companies has arguably been made clear, i.e. to foster the growth of the economy while shielding directors and shareholders of companies against full personal and unlimited liability, there is not an apparent reason requiring the creation of a *tertium genus* for smart autonomous surgical robots. Other approaches can bring about more opportune solutions.

It is important to note once again that robots are not companies. Robots (and surgical robots) are machines, instruments to diagnose or treat patients' illnesses.

---

[97]See Expert Group on Liability for New Technologies (2019), p. 61.

[98]See Expert Group on Liability for New Technologies (2019), p. 62.

Despite the significance of such a function, fundamental to peoples' health, its nature is entirely different from companies. Surgical robots do not need to be granted personhood, we do not need to debate about if somebody shall be obliged to pay social security on their behalf or a salary, or if they have the right to vacation, if they shall have right to vote, or if they have labor rights, or whether they have a right to rest. Shall their working time be limited to a certain number of hours per day? Do they have the right to unionize? These and other questions are not necessary, at least not in the present time. If this was the only available solution, we will be obliged to create a different status of personhood, in which the robot could be considered responsible, but without rights.

Finally, and to sum up, the solution of the EP proposed rules consisting in granting of personhood to autonomous robots is unnecessary as there exist other solutions that may deliver better and more efficient results, such as the solution of a mandatory insurance scheme for producers.

References

Alemzadeh H et al (2016) Adverse events in robotic surgery: a retrospective study of 14 years of FDA data. PLoS ONE 11(4):1–20. https://doi.org/10.1371/journal.pone.0151470

Alfaro Águila-Real J (2016) El reconocimiento de la personalidad jurídica en la construcción del Derecho de Sociedades. InDret (1)

Allen T, Widdison R (1996) Can computers make contracts? Harv J Law Technol 9(1):26–50

American Law Institute (2005) Restatement of the law third: agency, vol 1. American Law Institute Publishers, St. Paul

Aruni G et al (2018) New surgical robots on the horizon and the potential role of artificial intelligence. Investig Clin Urol 59:221–222. https://doi.org/10.4111/icu.2018.59.4.221

Bercovitz Rodríguez-Cano R (2001) Comentarios al Código Civil. Aranzadi, Elcano

Bertolini A (2016) Insurance and risk management for robotic devices. Global Jurist 16 (3):291–314. https://doi.org/10.1515/gj-2015-0021

Bertolini A et al (2016) On robots and insurance. Int J Soc Robot 8:381–391. https://doi.org/10.1007/s12369-016-0345-z

Brownsword R, Goodwin M (2012) Law and the technologies of the twenty-first century. Text and materials. Cambridge University Press, Cambridge

Cadiere G-B et al (2001) Feasibility of robotic laparoscopic surgery: 146 cases. World J Surg 25 (11):1467–1477. https://doi.org/10.1007/s00268-001-0132-2

Calo R (2011) Open robotics. Maryland Law Rev 70:101–142. http://ssrn.com/abstract=1706293

Calo R (2017) Artificial intelligence policy: a primer and roadmap. Soc Sci Res Netw:1–28. https://doi.org/10.2139/ssrn.3015350

Camarillo DB et al (2004) Robotic technology in surgery: past, present, and future. Am J Surg 188 (4A):2S–15S. https://doi.org/10.1016/j.amjsurg.2004.08.025

Cassarino A (2016) El papel de la actio tributoria en el sistema de los procedimientos concursales romanos. Rev Derecho Universidad Católica del Norte 23(2):243–264. https://doi.org/10.4067/S0718-97532016000200008

Chopra S, White LF (2011) A legal theory for autonomous artificial agents. The University of Michigan Press, Michigan

Delvaux M (2016) Draft Report with recommendations to the Commission on Civil Law Rules on Robotics (2015/2103/INL). Brussels

Delvaux M (2017) Report with recommendations to the Commission on Civil Law Rules on Robotics (2015/2103/INL). Brussels
Directorate-General for Internal Policies (2016) European civil law rules in robotics. Study for the JURI Committee. https://www.europarl.europa.eu/RegData/etudes/STUD/2016/571379/IPOL_STU(2016)571379_EN.pdf. Accessed 15 Apr 2020
Ebers M (2019) Chapter 2: regulating AI and robotics: ethical and legal challenges. In: Ebers M (ed) Algorithms and law. Cambridge University Press, Cambridge. https://doi.org/10.2139/ssrn.3392379
Ercilla García J (2018) Normas de derecho civil y robótica. Robots inteligentes, personalidad jurídica, responsabilidad civil y regulación. Thomson Reuters Aranzadi, Cizur Menor
European Commission (2018) Commission staff working document. Evaluation of Council Directive 85/374/EEC of 25 July 1985 on the approximation of the laws, regulations and administrative provisions of the Member States concerning liability for defective products. {COM (2018) 246 final} – {SWD(2018} 158 final}
Expert Group on Liability for New Technologies (2019) Liability for artificial intelligence and other emerging technologies. https://doi.org/10.2838/573689
Ganuza JJ et al (2016) Product liability versus reputation. J Law Econ Organ 32(2):213–241
Gray JC (1921) The nature and sources of the law. Columbia University Press, New York
Griffin A (2017) Facebook's artificial intelligence robot shut down after they start talking to each other in their own language. Independent. https://www.independent.co.uk/life-style/gadgets-and-tech/news/facebook-artificial-intelligence-ai-chatbot-new-language-research-openai-google-a7869706.html
Hansmann H et al (2006) Law and the rise of the Firm. Harvard John M. Olin Discussion Paper Series Vol. 2. No. 546
Holder C et al (2016a) Robotics and law: key legal and regulatory implications of the robotics age (part I of II). Comp Law Secur Rev 32(4):557–576. https://doi.org/10.1016/j.clsr.2016.05.011
Holder C et al (2016b) Robotics and law: key legal and regulatory implications of the robotics age (part II of II). Comp Law Secur Rev 32(4):557–576. https://doi.org/10.1016/j.clsr.2016.05.011
Houssami M et al (2017) Artificial intelligence for breast cancer screening: opportunity or hype? The Breast 36:31–33. https://doi.org/10.1016/j.breast.2017.09.003
International Organization for Standardization (2012) ISO 8373:2012
Intuitive Surgical (2018) Investor Presentation Q4 2018. https://isrg.intuitive.com/static-files/8afb7980-4820-41ff-bfa4-b3f82ce4111a
Intuitive Surgical (2019a) Investor Presentation Q2 2019. https://isrg.intuitive.com/static-files/0d3d24c1-849d-4495-8035-c17def42fd76
Intuitive Surgical (2019b) Investor Presentation Q4 2019. https://isrg.intuitive.com/static-files/0fc01a59-8d32-481f-9872-b262fd1f87b2
Jorrín JG (2019) El Pacto de Toledo abre la puerta a que los robots financien las pensiones. El Confidencial. https://www.elconfidencial.com/economia/2019-02-19/pacto-toledo-cotizacion-robots-seguridad-social_1833498/. Accessed 15 Apr 2020
Karanasiou A, Pinotsis D (2017) Towards a legal definition of machine intelligence. In: International Conference on Artificial Intelligence and Law (ICAIL), pp 1–10. London. https://doi.org/10.1145/3086512.3086524
Karnow CEA (1996) Liability for distributed artificial intelligences. Berkeley Technol Law J 11 (1):147–204. https://doi.org/10.15779/Z38ZD4W
Koops B-J et al (2010) Bridging the accountability gap: rights for new entities in the information society? Minn J Law 11(2):497–561
Lacruz Mantecón ML (2019) Cibernética y derecho europeo: ¿una inteligencia robótica? Diario La Ley, Sección Doctrina 9376:1–13
Leroux C et al (2012) Suggestion for a green paper on legal issues in robotics. Contribution to Deliverable D.3.2.1 on ELS Issues in Robotics

Moustris GP et al (2011) Evolution of autonomous and semi-autonomous robotic surgical systems: a review of the literature. Int J Med Rob Comp Assisted Surg 7(4):375–392. https://doi.org/10.1002/rcs.408
Muller C (2017) Artificial intelligence – the consequences of artificial intelligence on the (digital) single market, production, consumption, employment and society. Brussels. https://www.eesc.europa.eu/en/our-work/opinions-information-reports/opinions/artificial-intelligence. Accessed 15 Apr 2020
Muñoz Villarreal A, Gallego Corchero V (2019) Capítulo 2. Inteligencia artificial e irrupción de una nueva personalidad en nuestro ordenamiento jurídico ante la imputación de responsabilidad a los robots. In: Monterrosso Casado E (ed) Inteligencia artificial y riesgos cibernéticos. Responsabilidades y aseguramiento. Tirant lo Blanch, Valencia, pp 67–100
Nik-Ahd F et al (2019) Robotic urologic surgery: trends in litigation over the last decade. J Robot Surg 13(6):1–6. https://doi.org/10.1007/s11701-018-00905-y
Núñez Zorrilla MC (2019) Inteligencia artificial y responsabilidad civil. Reus Editorial, Madrid
O'Regan G (2016) Introduction to the history of computing. Springer International Publishing, Cham
O'Sullivan S et al (2019) Legal, regulatory, and ethical frameworks for development of standards in artificial intelligence (AI) and autonomous robotic surgery. Int J Med Robot Comp Assisted Surg 15(1):1–12. https://doi.org/10.1002/rcs.1968
Opfermann JD et al (2017) Semi - autonomous electrosurgery for tumor resection using a multi - degree of freedom electrosurgical tool and visual servoing. In: 2017 IEEE/RSJ International Conference on Intelligent Robots and Systems (IROS) 3653-60
Pagallo U (2013) The laws of robots. Springer, Dordrecht. https://doi.org/10.1007/978-94-007-6564-1
Pagallo U (2017) From automation to autonomous systems: a legal phenomenology with problems of accountability. In: IJCAI International Joint Conference on artificial intelligence, pp 17–23
Pagallo U (2018) Vital, Sophia, and Co. The quest for the legal personhood of robots. Information (Switzerland) 9(9):230. https://doi.org/10.3390/info9090230
Palmerini E et al (2014) Guidelines on regulating robotics. http://www.robolaw.eu/
Puaschunder JM (2018) Artificial intelligence evolution: on the virtue of killing in the artificial age. SSRN Electronic J. https://doi.org/10.2139/ssrn.3247401
Ramesh AN et al (2004) Artificial intelligence in medicine. Ann Royal Coll Surg England 86 (5):334–338. https://doi.org/10.1308/147870804290
Richards NM et al (2016) Robot law. Edward Elgar, Northampton
Rogel Vide C (2017) Personas, animales y androides. Revista general de legislación y jurisprudencia 4:681–693
Rogel Vide C (2018) Robots y personas. Revista general de legislación y jurisprudencia 2(1):81–92
Romero Otero J et al (2007) History, evolution and application of robotic surgery in urology. Archivos españoles de urología 60(4):335–341
Rusell S, Norving P (2008) Inteligencia artificial: un enfoque moderno. Pearson Educación, Madrid
Searle JR (1980) Minds, brains, and programs. Behav Brain Sci 3:417–457
Singer PW (2009) Wired for war: the robotics revolution and conflict in the 21st century. Penguin, London
Solum LB (1992) Legal personhood for artificial intelligences. North Carolina Law Rev 70 (9):1231–1287. https://doi.org/10.3868/s050-004-015-0003-8
Stapleton J (1989) Software, information and the concept of product. Tel Aviv Univ Stud Law 9:147–164
Stolfi E (2001) Studi sui «libri ad edictum» di Pomponio. Collana della Rivista di Diritto Romano, Milano
Strickland E (2017) In flesh-cutting task, autonomous robot surgeon beats human surgeons. IEEE Spectrum. https://spectrum.ieee.org/the-human-os/biomedical/devices/in-fleshcutting-task-autonomous-robot-surgeon-beats-human-surgeons. Accessed 15 Apr 2020
Surden H (2014) Machine learning and law. Wash Law Rev 89(1):87–115

Teubner G (2006) Rights of non-humans? Electronic agents and animals as new actors in politics and law. J Law Soc 33(4):497–521. https://doi.org/10.1111/j.1467-6478.2006.00368.x
Teubner G (2018) Digital personhood? The status of autonomous software agents in private law. Ancilla Iuris, pp 106–149. https://doi.org/10.2139/ssrn.3177096
The Headquarters for Japan's Economic Revitalization (2015) New robot strategy. Japan's robot strategy - Vision, strategy, action plan. https://www.meti.go.jp/english/press/2015/pdf/0123_01b.pdf. Accessed 15 Apr 2020
Wagner G (2019) Robot, Inc.: personhood for autonomous systems? Fordham Law Rev 88 (2):591–612
Wisskirchen G et al (2017) Artificial intelligence and robotics and their impact on the workplace. Report of the IBA Global Employment Institute (GEI), London. file:///C:/Users/irene/Documents/Estonia/Book/AI-and-Robotics-IBA-GEI-April-2017.pdf. Accessed 15 Apr 2020
Wolf A, Shoham M (2009) 78. Medical automation and robotics. In: Nof SY (ed) Springer handbook of automation. Springer-Verlag, Berlin, pp 1397–1407
Yampolskiy RV (2018) Could an artificial intelligence be considered a person under the law? The Conversation. http://theconversation.com/could-an-artificial-intelligence-be-considered-a-person-under-the-law-102865. Accessed 15 Apr 2020

Online Behavioural Advertising and Unfair Manipulation Between the GDPR and the UCPD

Federico Galli

Contents

F. Galli (✉)
CIRSFID, Bologna, Italy
e-mail: federico.galli7@unibo.it

M. Ebers, M. Cantero Gamito (eds.), *Algorithmic Governance and Governance of Algorithms*, Data Science, Machine Intelligence, and Law 1,
https://doi.org/10.1007/978-3-030-50559-2_6

Abstract Online behavioural advertising, i.e. the practice of targeting digital advertising according to consumers' online behaviour, is posing threats to consumers' autonomy. This is happening not only because behavioural advertising intrinsically has the effect of restricting options for consumers, but also and especially because it increasingly enables customisation of advertising messages according to psychological traits and vulnerabilities of consumers. Against this background, the present work evaluates if, how and to what extent the EU General Data Protection Regulation EU/2016/697 (GDPR) and the Directive 2005/29/EC on unfair commercial practices (UCPD) contribute to safeguarding autonomous decision-making of individuals in front of new data and technology-driven subtle forms of manipulative advertising.

1 Introduction: Online Behavioural Advertising and Unfair Manipulation

Over the past few years, the convergence between Big Data and machine learning techniques has profoundly changed the pervasiveness and effectiveness of online behavioural advertising (also "OBA"). With the exponentially growing amount and variety of consumers' data and the ubiquitous employment of learning algorithms, today marketers are able to develop extremely effective advertising strategies by figuring out which ads each consumer best responds to.

These advancements are generally presented to consumers as an added value. Admittedly, consumers nowadays face an overload of potential commercial deals. By highlighting the relevance of goods and services, which consumers might be more interested in, OBA decreases search and transaction costs, and maximise the satisfaction consumers gain from their choices.[1]

Nevertheless, concerns about consumer welfare have also been voiced. Besides the well-known problem of "autonomy trap"[2] and "filter bubble",[3] many authors have highlighted the subtle and powerful nature of "nudges" that are currently deployed by online advertisers to shape the informational context where consumers make decisions and to steer them to make predetermined consumption choices, e.g. ranging from the order of search results to the specific methods with which a message is addressed to consumers.[4] In addition to that, contributions from different

[1] Noto La Diega (2018), p. 53; Matz et al. (2016).

[2] Schwartz (2000) and Zarsky (2003).

[3] Pariser (2011).

[4] Weinmann et al. (2016), Yeung (2017) and Susser (2019).

fields of study highlight how novel and increased information asymmetries between marketers and consumers enable and even encourage advertisers to exploit consumers psychological vulnerabilities by tailoring advertising to their psychological traits and vulnerabilities.[5]

Both issues clearly point in the same direction: a scenario where companies track and profile consumers, study their behaviour and use their digital footprint as laboratories to develop for each consumer the persuasive technique that proves to be most effective in terms of clicks and conversions. With the spread of wearable devices, the growing uptake of the Internet of Things ('IoT'), and the related flood of contextual and particularised data,[6] authors warn that marketers will be increasingly able to cover and trigger consumer frailty at individual level, at any place and at any time.[7]

Against this background, a question has been raised about the "legally permissible levels of transactional exploitation".[8] The question is not easy to solve because advertisers have always found ways to influence consumers' decision-making and to push them to make specific purchase decisions.[9] Accordingly, the line between "informing, nudging and outright manipulation" has been difficult to draw by legislators,[10] both because of the risk of adopting a too paternalistic view and because of the constant ability of companies to resort to new and ever-changing marketing techniques. Yet, we might expect that this is the task of our legislators if they care to protect their citizens vis-à-vis new and powerful data-driven, potentially manipulative advertising techniques.

Along these lines, the purpose of this chapter is to analyse (1) under what conditions online behavioural advertising is allowed in the existing EU legal framework and (2) whether these conditions can avoid or control the risk of consumers' manipulation in the era of "data-driven advertising".

To this aim, I will consider the EU General Data Protection Regulation (GDPR) and the Directive on Unfair Commercial Practices (UCPD), as the most representative regulatory tools from the respective fields of European data and consumer protection law. Both of these fields are concerned with the protection of the autonomy of natural persons, notably of data subjects, on the one side, and consumers, on the other side. Data protection law affords protection to individual autonomy more in a moral sense and for human rights protection-oriented reasons; consumer protection law for market-related purposes.[11] However, in the current digital economy "these worlds start coming together".[12] Consumption has largely

[5]Bar-Gill (2013), Calo (2013), Clifford (2017) and Matz et al. (2017).

[6]Helberger (2016).

[7]Calo (2013).

[8]Mik (2016).

[9]Hanson and Kysar (1999).

[10]Helberger (2016).

[11]Nissenbaum (2010), pp. 81–84.

[12]Helberger et al. (2017).

become digital and consumers are generally also data subjects that pay for free content with their personal information.[13] In front of targeted advertising, individuals are not merely data subjects whose personal data is being processed and profiled, but potential consumers that may be "gently pushed" into a transactional decision. At the same time, data processing is abundantly taking place on the Internet for advertising purposes. The ad industry aggregates nearly all the web traffic in the form of ads transactions across websites and users globally.[14] Thus, it seems that data protection policy can no longer neglect its most critical *commercial* battlefield, i.e. the market of online advertising, in which data represents the commodity.[15] That is why, throughout the next sections, I will refer to "consumer-data subject" as to denote the person that faces new advertising practices.

It is nevertheless worth pointing out that data protection and consumer law have different stories and still show different important characteristics.[16] Above all, "consumer law deals with fair contracting; data protection law with fair processing".[17]

The chapter is organised as follows. Section 2 provides an overview of the GDPR relevant provisions that apply to online behavioural advertising and that can abstractly provide online consumers with safeguards against behavioural advertising manipulative practices. With a similar approach, Sect. 3 focuses on the other tier of our analysis, i.e. the UCPD. Section 4 concludes by comparing the two legislative frameworks to outline how differently the two frameworks tackle the problem of manipulation of consumer data subjects. The work ultimately argues that the UCPD comparatively represents a better groundwork for protecting consumers in the era of data-driven and technology-mediated influence, thereby paving the way for further discussion.

2 Online Behavioural Advertising, Unfair Manipulation and the GDPR

The GDPR does not explicitly address online behavioural advertising, nor does it tackle directly the problem of manipulation of data subjects. Nonetheless, OBA is relevant under the GDPR for two critical reasons:

---

[13]For a combined legal analysis of consumers and data subjects, as well as the integration between the relevant fields of law, see Schmechel (2017) and Micklitz (2018).

[14]Recent statistics on the state of online advertising industry in EU and relevant usage can be found at https://ec.europa.eu/eurostat/statistics-explained/index.php/Internet_advertising_of_businesses_-_statistics_on_usage_of_ads#General_overview. Accessed 3 Dec 2019.

[15]Noto La Diega (2018), pp. 445–499.

[16]Rhoen (2016).

[17]Helberger et al. (2017), p. 1428.

1. First of all, the practice of targeting advertising according to consumers' behaviour is a complex process that entails different activities of personal data processing that are carried out by multiple commercial entities.[18] It follows that each data processing operation must be carried out complying with the principles and provisions of the Regulation. At the same time, from the perspective of the GDPR, behavioural advertising as such is one of the most frequent referred purposes for the processing of personal data.[19]
2. Secondly, OBA represents the most ubiquitous example of automated decision-making, that typically includes profiling ("any form of automated processing of personal data consisting of the use of personal data to evaluate certain personal aspects relating to a natural person, in particular to analyse or predict aspects concerning that natural person's performance at work, economic situation, health, personal preferences, interests, reliability, behaviour, location or movements").[20] In this regard, Article 22 GDPR prohibits decisions based on solely automated processing that have a legal or a similarly significant effect on data subjects and leave doubt as to whether the prohibition also applies to OBA.

2.1 OBA As a Business Practice Entailing Different Processing Activities: Relevant Legal Basis and Overarching Principles

OBA in itself is not a processing activity, but a commercial practice that entails different kinds of processing activities. Following the scheme adopted by Zuiderveen Borgesius,[21] online behavioural advertising unfolds in five phases: (1) *data collection* is the activity of collecting information about people's online behaviour. This generally happens through the use of cookies or other tracking technologies (e.g. web beacons, device fingerprinting, software installation); (2) *data storage*, i.e. the recording of the data tied to a unique identifier; (3) *data analysis* to extract useful information from consumers' data through mining techniques; (4) *data disclosure* consists in making data available to other firms than the one that collected it; and (5) *targeting advertising* that is the automated delivery of advertising on the user's web page.

The process involves several subjects. Marketing companies and website owners represent demand and supply in the market of ad spaces. These are connected by a

---

[18] Although not linked to personal names, it is agreed that behavioural data processed in the course of OBA are to be generally considered personal data, given their capacity to single out a natural person. See Article 29 Working Party (2010a), p. 9. For a detailed discussion, see Zuiderveen Borgesius (2016).

[19] Contissa et al. (2018).

[20] Article 4(4) GDPR.

[21] Zuiderveen Borgesius (2015), p. 53.

number of digital platforms, each one playing a different role in the targeting process. Among these, *supply-side platforms* (SSDs) work on behalf of publishers and offer them optimised strategies to manage their advertising inventory; *ad exchanges* (ADXs) aggregate ad spaces from different publishers and sell them to advertisers by means of auctions based on real-time bidding technology; *demand-side platforms* (DSPs) serve marketing companies by automatically bidding for the campaigns in multiple ad exchanges. Each platform can set its cookies on different websites to collect as much consumer information as possible and employ various data analysis techniques (sometimes advanced, such as in the case of machine learning) to extract useful knowledge. Collection and analysis are otherwise performed by separate subjects, such as *data management platforms* (DMPs) or *data exchanges platforms* (DXPs), that directly sell analysed data in the form of profiles or predictive scores to the platforms that are involved in the ad delivery. We will refer to all these different entities as "ad platforms".[22]

2.1.1 Legal Basis for Processing

When processing personal data, ad platforms generally act as data controllers and must comply with the provisions of the GDPR.[23] In particular, any data processing activity needs to have a clear legal basis among those enlisted in Article 6.

Consent and legitimate interest are particularly relevant in the context of OBA.[24] In 2010, the WP29 had clarified that, under the Directive 95/46/EC, OBA always needed data subject's consent as a legal basis.[25] This notwithstanding, the GDPR explicitly introduced in Recital 47 that "processing of personal data for direct marketing purposes may be regarded as carried out for a legitimate interest", apparently disavowing the WP29.[26] This inconsistency has brought up the question as to whether OBA shall be based on consent or a legitimate interest of the data controller. The question is not trivial because consent relies on an opt-in scheme, whereas legitimate interest reserves the data subject the right to object when the

[22] A detailed explanation of the technical functioning of RTB and the relevant actors is contained in Wang et al. (2017).

[23] See Article 29 Working Party (2010a). The 29WP here clarifies the obligations of each actor involved in OBA under the framework of Data Protection Directive.

[24] Contract as a legal basis for the processing has been ruled out by 29WP, see Article 29 Working Party (2014), p. 17. See also Article 29 Working Party (2013), p. 46. Cfr. with Zuiderveen Borgesius (2015).

[25] Article 29 Working Party (2010a).

[26] Online behavioural targeting can be seen as a type of direct marketing as confirmed in the code of conduct of the Federation of European Direct and Interactive Marketing: "Direct marketing in the online environment refers to one-to-one marketing activities where individuals are targeted". The Working Party approved the code in Article 29 Working Party (2010b).

processing is based on direct marketing purposes (Article 21(2) GDPR). This remains a debated topic, but it is not a specific object of this contribution.[27]

Consent is not only one of the legal bases for processing personal data according to Article 6. It is emblematic in the context of OBA for other three reasons:

(a) When setting cookies to collect information about data subjects, be that personal or non-personal data, ad platforms must comply with requirements of notice and consent, as set out in Article 5(3) ePrivacy Directive 2002/58/EC, as amended by the Directive 2009/136/EC,[28] and as interpreted by the EU Court of Justice in the recent *Planet 49* judgment.[29]
(b) Explicit consent is one of the grounds of exemption of the prohibition of Article 22 GDPR (see Sect. 2.2).
(c) Explicit consent is also meaningful for Article 9 GDPR. Commercial players involved in advertising targeting may process special categories of data. In such a case, the legal grounds for processing are more limited and the legitimate interest of the data controller is not included. Of course, it will be difficult to foretell whether personal data used in OBA may constitute special categories of data and need to be processed according to Article 9. This occurrence will depend on the specific data content (health, political opinion etc.) and admittedly, data used for behavioural targeting does not generally reveal a consumer's sensitive attributes. However, data analysis techniques can create special categories of data from data which is not a special category data in its own right but becomes so when combined with other data. For example, it may be possible to infer someone's state of health from the records of their food shopping combined with data on the quality and energy content of foods.[30] As noted by Clifford, with the rise of machine learning methods, the lines between personal and sensitive personal data have become increasingly blurred.[31] This is even

---

[27]In any case, one must consider that *legitimate interest* does not offer a free card to ad platforms for the processing of personal data for direct marketing purposes. A balancing test between the interest of the controller and the data subjects' fundamental rights and freedom is required, see Zuiderveen Borgesius (2015), p. 168.

[28]Directive 2002/58/EC of the European Parliament and of the Council of 12 July 2002 concerning the processing of personal data and the protection of privacy in the electronic communications sector (Directive on privacy and electronic communications), last amended by Directive 2009/136/EC of the European Parliament and of the Council of 25 November 2009 (OJ L 337 11.). Consider, however, that the consent requirement for the placement of cookies is regarded as being different from the consent for the processing of personal data within the meaning of the GDPR, see Zuiderveen Borgesius (2015).

[29]Judgment of the Court (Grand Chamber) of 1 October 2019 Bundesverband der Verbraucherzentralen und Verbraucherverbände—Verbraucherzentrale Bundesverband eV v Planet49 GmbH, C-673/17, EU:C:2019:801. In the judgment, the Court stated that consent is not validly constituted if the storage of information, or access to information already stored in the website user's terminal equipment, is permitted by way of a pre-ticked checkbox which the user must deselect to refuse his consent.

[30]See Article 29 Working Party (2017), p. 15.

[31]Clifford (2017), p. 25.

more true if one considers that information commonly used for the most fastidious and manipulative forms of digital advertising is related to psychological traits and emotional states of consumers. Such information is neither sensitive nor personal data but is generally the result of statistical inferences on behavioural and social media data, such as time spent on-site or likes on Facebook.[32]

2.1.2 Principles

Besides specific provisions, the GDPR lays down the principles that must govern any data processing activity, thus also the operations resulting in online behavioural advertising. With regard to the risk of manipulation, the principles of lawfulness and fairness may play a critical role in the assessment of the GDPR-compliance of data-driven advertising techniques.

The GDPR requires that personal data are processed *lawfully*, fairly and in a transparent manner.[33] The principle of lawfulness of processing implies that data is to be processed by complying with all the provisions of the GDPR and with all regulations that are issued on its basis. It is directly connected to Article 6 that sets out the grounds for lawful processing. According to Clifford and Ausloos, the principle of lawfulness must be however distinguished by the inherent principle of legality, which, on the contrary, stems from the principle of purpose limitation of Article 5(1)(b).[34] According to the latter, personal data shall be collected for specified, explicit and *legitimate purposes* and not further processed in a way incompatible with those purposes.[35] It follows that an evaluation of the legitimacy of the purposes of the processing would therefore represent a preliminary step for assessing the lawfulness of the processing operation.

With regard to data-driven advertising practices, even before assessing the presence of a lawful basis for processing (see Sect. 2.1.1), it would be necessary to understand whether: (1) beyond statements made on paper, the purpose of the processing is that of deliberately subverting consumer data subjects' decision-making (such as in the case of intentional manipulation); (2) this purpose entails an activity that is considered illegal according to other EU laws (such as in the case of Unfair Commercial Practice Directive, see Sect. 3). In the latter case, not only would the practice be prohibited under other relevant laws, but it would be unlawful and prohibited also according the GDPR.

Article 5 also requires that data controllers process personal data *fairly*. At a very general level, fair processing of personal data is interpreted as processing that respects the interests and reasonable expectations of data subjects.[36] More

[32]Burr and Cristianini (2019).

[33]Article 5(1)(a) GDPR.

[34]Clifford and Ausloos (2018), p. 162.

[35]Article 5(1)(a) GDPR.

[36]Bygrave (2002), p. 58.

specifically, Clifford and Ausloos argue that fairness in the GDPR would have an explicit and implicit dimension. Explicit fairness refers to the procedural regularity that data controllers are required to follow during the entire processing. The implicit dimension of fairness instead requires a fair balancing test between the data controller's commercial interest and the data subject's rights. This operation is made by the GDPR itself, both before (i.e. conditions for lawful data processing) and after the processing (i.e. data subject rights).[37] Under this second meaning, the principle of fairness might represent an implicit safety net to protect consumer data subjects against manipulative advertising: personal data should not be processed *in a way* that disproportionately infringes the fundamental rights and freedoms of data subjects. Therefore, fairness may open up to consider not only how data is processed, but also the *effect* that the processing has on data subjects. But the fact remains that the broadness and vagueness of fairness in the GDPR make it still very difficult to grasp its precise meaning. That is why it has been argued that the data protection principle of fairness should be inspired by the elaboration of the principle of fairness in consumer protection law.[38]

2.2 OBA As "Automated Decision-Making, Including Profiling"

Another provision that is relevant for the regulation of online behavioural advertising is Article 22. The latter establishes that "the data subject shall have the right not to be subject to a decision based solely on automated processing, including profiling, which produces legal effect concerning him or her or similarly significantly affects him or her".[39]

As pointed out by Veale and Edwards, behavioural advertising represents one of the most pervasive online experiences of *solely* automated decision-making.[40] With the real-time bidding ecosystem, OBA takes place within 100 milliseconds. No human could ever be involved in the process without the cost of making the entire operation totally inefficient and ineffective for its purpose.

It remains, however, far from clear whether behavioural targeting entails a *decision* and more importantly, whether it can produce a legal effect or *similarly significantly affects* the consumer data subject.

[37] Clifford and Ausloos (2018), p. 144.

[38] Helberger et al. (2017) and Graef et al. (2018).

[39] Article 22(1) GDPR.

[40] Veale and Edwards (2018), p. 401.

2.2.1 Decision

Following Mendoza and Bygrave, for a *decision* to be made, a particular attitude or stance must be taken towards the data subject and this should have some degree of binding effect in the sense that is likely to acted upon.[41] Advertising has been defined as any "paid, mediated form of communication from an identifiable source, designed to persuade the receiver to take some action, now or in the future".[42] Whereas usually the purpose of other traditional forms of advertising is to raise the brand visibility and salience with a long-term effect (sc. branding advertising), today's digital advertising intends to entice the recipient to take action by immediately clicking (sc. direct response marketing).[43] So, it seems that OBA is by definition a stance that is likely to be acted upon immediately. In any case, a decision in itself needs not to be binding in a legal sense, otherwise it would be hard to grasp the logic as to why Article 22 prohibits the decision-making only when the *decision has a legal effect* (or similarly significant effect).

2.2.2 Similarly Significantly Affects

While it is clear that even in its most subtle and manipulative forms, advertising does not have a legal effect (advertising is not a command and can be ignored or even blocked with the use of ad-blocking tools), we still need to assess whether and when online behavioural advertising is able to *significantly affect* consumer data subjects in a similar way. In this case too, OBA would be prohibited.

In the Guidelines on Automated individual decision-making and Profiling, 29WP explained that the threshold for significance must be similar to that of a decision producing a legal effect and "the effects of the processing must be *sufficiently great or important to be worthy of attention*".[44] Three factors are enlisted that can determine where the effect is significant.

A decision must have the potential to:

- significantly affect the circumstances, behaviour or choices of the individuals concerned;
- have a prolonged or permanent impact on the data subject;
- or at its most extreme, lead to the exclusion or discrimination of individuals.[45]

The first parameter seems rather significant for our purpose. Circumstances, behaviour and choice are recurring terms among the theorists of "libertarian paternalism"[46] or "nudging".[47] In their view, society should exploit irrational human

[41] Mendoza and Bygrave (2017), p. 87.

[42] Richards and Curran (2002), p. 63.

[43] McStay (2016).

[44] Article 29 Working Party (2017), p. 21.

[45] Article 29 Working Party (2017), p. 21.

[46] Thaler and Sunstein (2003).

[47] Thaler and Sunstein (2009).

tendencies in order to shape context where citizens make choices to drive them toward preferred behaviour, while still leaving individuals technically capable of resisting external intervention. The problem of digital market manipulation that arises with data-driven advertising has been also described as "private nudging" or "nudging for profit":[48] companies use what they know or infer about consumers' psychology to influence their decisions in the online context.[49] Authors have also highlighted that in digital environments the problem of "choice architecture" is even more critical, given the strategically-designed omnipresence of nudges in the Web[50] and their capacity to adapt to each user.[51] Considering these claims, one could argue that behavioural advertising does have at least the *potential* to significantly affect the circumstances, behaviour or choices of the individual concerned and thus may be of sufficient importance to be worthy of attention.

Interestingly, WP29 itself explicitly refers to online behavioural advertising.[52] According to the Authority, OBA generally does not have a similarly significant effect on individuals. The example here provided (i.e. targeting based on profiles such as "women in the Brussels region aged between 25 and 35 who are likely to be interested in fashion and certain clothing items") seems quite simplistic compared to the accuracy and the intrusiveness of the knowledge that can be extracted from simple behavioural data.[53] However, following WP29:

> some particular forms of OBA may do [have a similarly significant effect on individuals] depending on the particular characteristics of the case, including:
>
> - the intrusiveness of the profiling process, including the tracking of individuals across different websites, devices and services;
> - the expectations and wishes of the individuals concerned;
> - the way the advert is delivered;
> - or using knowledge of the vulnerabilities of the data subjects targeted.[54]

Based on this, one can distinguish between a "soft" and "strong" OBA, that are subject to different regimes. *Soft OBA* will not be subject to the prohibition of Article 22 GDPR.[55] On the other hand, *strong OBA* "may" qualify as automated decision-making that has a similarly significant effect on data subjects depending on: the

[48]Calo (2013), p. 1001.

[49]Weinmann et al. (2016) define "digital nudging" as "*the use of user interface design elements to guide people's choice or influence users' inputs in online decision environment*".

[50]Susser (2019).

[51]Yeung (2017).

[52]See Article 29 Working Party (2017), p. 22 ("*This brings us also to the issue of online advertising, which increasingly relies on automated tools and involves solely automated individual decision-making*").

[53]Matz and Netzer (2017).

[54]Matz and Netzer (2017).

[55]Types of behavioural advertising that do not have a significant similar effect on data subjects still entail "profiling" in the meaning of Article 4(4) GDPR. Therefore, even though not subject to Article 22, they are regulated under Article 13 and 14, which mandate a set of information that must be disclosed to the data subject.

intrusiveness of data collection techniques; the peculiarities of the consumer's context; the advertising practice and the exploitation of the data subject's personal vulnerabilities. In such cases, OBA would be prohibited under Article 22 GDPR.[56]

2.2.3 Exemptions

From the authoritative interpretation of WP29 on Article 22 GDPR, there are reasons to think that, if the indications of the Authority were to be followed to the letter, new manipulative advertising practices that are described under the label "digital market manipulation" would be prohibited under Article 22 GDPR.[57]

If that is the case, Article 22(2) becomes noticeably relevant because it provides some exemptions to the prohibition contained in Par. 1. This means that the data controller must not carry out an automated decision-making, including profiling, that falls within the prohibition of Article 22(1), unless the three exemptions of Article 22 (2) apply. These refer to the circumstances where the decision is: (1) necessary for the performance of or entering into a contract; (2) authorised by Union or Member States law to which the controller is subject and which also lays down suitable measures to safeguard the data subject's rights and freedoms and legitimate interests; or (3) based on the data subject's explicit consent.

All three conditions will not be easily satisfied in the context of behavioural advertising. First, advertising may be well the means thank to which consumers enter into a contract: without online advertising, the consumer could not often access a website free-of-charge. In such cases, an exchange of value is therefore implied: access to free content in exchange for consent of the user to be targeted with advertising. If this is the case, however, the controller must be able to show that automated decision-making is *necessary*, particularly considering whether a less privacy-intrusive method could be adopted. Other forms of online advertising, e.g. contextual advertising, are possible without tracking people's behaviour across multiple websites. It appears that OBA is not strictly necessary for the conclusion of a contract, but it is needed to maximise the chance that one consumer enters into a contract, therefore the profitability of the deal for advertisers. As for the second exemptions, it would be entirely unexpected if Union or Member State law allowed companies under their jurisdiction to perform the most intrusive forms of automated decision-making, accepting the risk of manipulation of their national consumers.

As for the third exemption, the GDPR requires that, to be explicit, consent must reflect a higher standard of protection than the one enshrined in the "normal" consent of Article 4(11) GDPR. The term explicit refers to the way consent is expressed by

[56]Similarly, the UK Data Protection Officer [ICO (2017)] recognised that, "even though many organisations might believe that advertising does not generally have a significant adverse effect on people, some studies proved that behaviourally targeted adverts can have psychological consequences and affect individuals self-perception".

[57]It has been noted that the guidance, despite some ambiguities in wording, resembles more unauthorised law-making than mere interpretation. See Veale and Edwards (2018), p. 401.

the data subject through an express statement. In an online context, this could be done, for example, "by offering an *explicit consent screen* that contains 'Yes' and 'No' checkboxes, provided that the text clearly indicates the consent".[58]

As seen in the previous section, consent already represents the legal basis for lawful processing in the context of OBA. To what extent this, in practice, falls under the requirements of "explicit consent" too, is rather doubtful. One can imagine that in real-time bidding, which fundamentally relies on a behind-the-scenes process, it is hard for every company that is responsible for creating profiles and displaying the ads to obtain explicit consent, which in turn must be specifically related to the automated decision-making. Therefore, it appears urgent to carry out further research in order to shed light on to what extent the behavioural advertising industry is complying with GDPR and Article 22.[59]

What seems an absolute priority, is a more specific and technology-aware delineation of the practices that make behavioural advertising have a similarly significant effect on data subject (*strong OBA*). This step, however, should arguably take a holistic approach and also consider protection standards in the domain of consumer law.

3 Online Behavioural Advertising, Unfair Manipulation and the UCPD

Similarly to the GDPR, the Directive on Unfair Commercial Practices does not touch upon online behavioural advertising.[60] Its provisions are nevertheless relevant since they regulate all commercial practices between traders and consumers, including advertising and marketing.[61]

---

[58] Article 29 Working Party (2011), p. 18.

[59] In any case, it must be recalled that even if processing activities based on automated decision-making were permissible under one of the exceptions provided for in Art. 22(2), in particular, "explicit consent", they would be subject to suitable safeguards to protect the data subject's rights and freedoms and legitimate interests. These are provided in Article 22(4) and include specific information to the data subject, the right to obtain human intervention on the part of the controller, to express its point of view, to obtain an explanation of the decision reached after such assessment and to contest the decision. Nonetheless, it seems that none of these appear to be useful and consistent if both the specificity of the OBA practice and the problem of manipulative advertising are considered.

[60] Online behavioural advertising is only mentioned in the 2016 Commission's Guidance of the UCPD: see European Commission (2016a), p. 134 and note 310. Online behavioural advertising is introduced in the Guidance together with other commercial practices that are typical of the algorithmic age and entail personalisation, e.g. dynamic pricing, price discrimination, personalised pricing, but is not further discussed.

[61] Article 2(d) UCPD, defines "business-to-consumer commercial practice" as "any act, omission, course of conduct or representation, commercial communication including advertising and marketing, by a trader, directly connected with the promotion, sale or supply of a product to consumers". It

More specifically, the UCPD deserves a special attention because it is the primary legal instrument within the EU that addresses manipulative commercial practices. Indeed, Article 5 prohibits "unfair commercial practices", i.e. those practices that, contrary to the requirement of professional diligence, materially distort or are able to materially distort the economic behaviour of the average consumer they reach.[62] Besides this, the sc. "general clause" of unfairness, the Directive also prohibits misleading actions/omissions[63] and aggressive practices (Sect. 3.1). Finally, it stipulates a blacklist of practices that are unfair *per se* and therefore prohibited irrespective of the circumstances. Therefore, when a practice is not part of the blacklist and does not fall under the definition of misleading and aggressive practices, it can be still considered unfair if it falls under the general clause of unfair commercial practices. The rationale of the UCPD is to stimulate business to advertise and market across borders and to stimulate consumers to actually make cross-border transactions thus contributing to the completion of the internal market.[64] It is therefore important to acknowledge that the Directive is not only concerned with eliminating distortion of the consumer's decision-making as such, but also and above all because this will result in the removal of distortion of the competition.[65]

To achieve this, the UCPD seeks to strike a complicated balance between empowering and encouraging the consumer to make autonomous decisions through the information paradigm, and protecting consumers from situations in which they are unable to protect themselves, either because they are in a position of vulnerability or because of the particular unfairness of the practice.[66]

The following sections address the question as to what extent the provisions of the UCPD can help protect consumer data subjects against behavioural advertising.

3.1 OBA As "Aggressive Commercial Practice"

Article 8 UCPD establishes that a commercial practice is aggressive "*if, in its factual context, taking account of all its features and circumstances, by harassment, coercion, including the use of physical force, or undue influence, it significantly impairs or is likely to significantly impair the average consumers freedom of choice or conduct with regard to the product and thereby causes him or is likely to cause*

is not necessary that the trader who undertakes the commercial practice is the one who then supplies the product to the consumers; this can be someone further below in the supply chain. See Law Commission (2004), p. 21 and note 24.

[62] Article 5(2) UCPD.

[63] Article 6 and 7 UCPD.

[64] Stuyck et al. (2006).

[65] As explained by Stuyck et al. (2006), "traders acting unfairly will no longer be in a position to win consumers away from competitors who play by the rules".

[66] Sax et al. (2018), p. 104.

him to take a transactional decision that he would not have taken otherwise".[67] As noted by Helberger, "it is easy to see its relevance [. . .] for the matter of 'digital market manipulation'".[68]

Let us imagine the case of Alice: she is a great fan of travelling to the Far East and registered to an online meditation course. Last week Alice bought a new pillow and mattress on an e-commerce website. She also visited two blogs on relaxing Indian herbal tea leaves. Alice usually stays up late surfing online. She never goes offline before 3 a.m. Thursday, at 4 in the morning, she cannot sleep and starts playing her favourite online game. Every time she gets through the next level, she is targeted with a piece of advertising that offers a discount on a particular brand of homoeopathic sleeping pills that can reduce insomnia in only 4 weeks. She is nervous and tired. She has not been sleeping for long. She clicks on the ad and buys two packs. Is this a crafty marketing strategy or an aggressive commercial practice?

As explained by Howells, aggressive commercial practices represent the clearest examples wherein business conduct impinges the most on consumer's freedom to choices.[69] It is nonetheless difficult to draw a line between what constitutes an aggressive practice and what remains legitimate persuasion, which is the ultimate aim of advertising. To establish that, Article 8 UCPD requires three conditions to be met.

3.1.1 The Practice Must Harass, Coerce or Unduly Influence Its Addressee

In establishing whether a commercial practice uses harassment, coercion, or undue influence, Article 9 UCPD specifies that one must consider inter alia the timing, the location, nature and persistence of the practice, and also "*the exploitation by the trader of any specific misfortune or circumstance of such gravity as to impair the consumers judgement, of which the trader is aware, to influence the consumers decision with regard to the product*". In the context of OBA, all three elements of Article 8 may be of relevance.

3.1.1.1 Harassment

Harassment has been defined as "pursuing, bombarding or pestering the consumer with inconveniences or requests".[70] Online consumers can be reached by targeted advertising multiple times, anywhere and at any time when they are surfing on the

[67] Article 8 UCPD.

[68] Helberger (2016).

[69] Howells et al. (2016), pp. 167–194.

[70] Carballo-Calero (2016).

Internet (at any time, in absolute, when consumers interact with wearable and IoT devices).[71] Interestingly, in evaluating a practice as harassment, the invasion of privacy is considered one of the most relevant factors.[72] In the example of Alice, the combination of multiple targeting at her home at 3 a.m. when she is predictably tired may be considered harassment.

3.1.1.2 Undue Influence

Undue influence is defined by the Directive as "exploiting a position of power in relation to the consumer so as to apply pressure, even without using or threatening to use physical force, in a way which significantly limits the ability of consumers to make an informed decision".[73] Digital advertisers and online consumers might find themselves in an imbalanced commercial relationship. Advertisers, or those operating on their behalf, have detailed knowledge about consumers' behaviours, preferences, needs and even fears, misperceptions and emotions, more than consumers themselves. In turn, consumers know nothing about who is behind their screen targeting advertising. This different informative power enables traders to adapt the right persuasion strategy to each consumer. If they effectively do so, there could be the case of exploitation of a position of power amounting to "undue influence".[74] In our example, the practice of targeting advertising on sleeping pills clearly leverages on the ability the trader has to profile Alice as presumably suffering from insomnia.

3.1.1.3 Coercion

Coercion represents all the residual cases in which harassment and undue influence do not apply. Not a physical, but more psychological coercion could be a trigger to prohibit manipulative advertising techniques, even when an advertiser does not retain a position of power over consumers. Playing on emotions and the exploitation of trust in consumers might need to be reviewed under this standard.

---

[71] Helberger (2016).

[72] For instance, conducting personal visits to the consumers home ignoring the consumers request to leave (Annex I, no. 25 UCPD), or making persistent and unwanted solicitations by telephone, fax, e-mail or other remote media (Annex I, no. 26 UCPD).

[73] Article 2(j) UCPD.

[74] The question here is whether this circumstance, as Howells put it, "creates an imbalance that is more than the general advantages traders have over consumers", see Howells et al. (2016), p. 188. Only in such a case, the "position of power" can be relevant enough to lead to an unfair influence.

3.1.2 The Practice Must Impair or Likely Impair Significantly the Consumer's Freedom of Choice

To be relevant under Article 8, harassment, coercion, and undue influence must significantly impair or likely impair the average consumer's freedom of choice or conduct. The fact that the average consumer is taken as the standard suggests that not every practice that makes consumers feel uncomfortable or pressured will be punishable (see Sect. 3.2.1). According to Helberger, "under which conditions profiling and targeting can amount to significantly impairing or being likely to impair a consumer's freedom of choice or conduct significantly will depend on: a) the *persuasive potential of personalised advertising*; and b) the extent to which the practice *reduces the autonomous decision-making process*".[75] As for the first, we can refer to Kaptein et al. who present "persuasion profiling" as a method to personalise the persuasive messages used by a system to influence its users.[76] Also, Burr et al. discuss the ethical, social and legal implication of persuasive technologies and argue that they can exploit and reinforce weaknesses in human beings.[77] Regarding the second, an emerging field of research is investigating how behavioural advertising affects individual consumer's choices.[78] Obviously, the extent to which the option for consumers to ignore the advertising and remove themselves from the trader's influence is feasible and practical (for instance, through the use of ad-blocking software) shall also be considered to evaluate the impairment on consumers' freedom of choice.

3.1.3 The Practice Must Cause or Likely Cause the Consumer to Take a Transactional Decision He Would Not Have Taken Otherwise

It is not enough that the consumer's freedom is impaired. This must in turn cause or be likely to cause him/her to take a transactional decision that he/she would not have taken otherwise. In the case of OBA, advertising is personalised to the individual situation and preferences. Helberger argues that in such cases, the likelihood that the consumer already had a predicament for taking that particular decision is higher.[79] However, there is a thin line between showing consumers what they need, what they think they need and showing what they fear they need. Most manipulative advertising techniques might be so invasive that they are able to profile consumers even taking into specific account their bias and fears. If targeted at the right moment and at the right time, behavioural advertising can presumably alter specific consumer transactional decisions.

---

[75] Helberger (2016).

[76] Kaptein et al. (2015).

[77] Burr et al. (2018).

[78] Bright and Daugherty (2012) and Souiden et al. (2017).

[79] Helberger (2016).

3.2 OBA, the Rational and the Vulnerable Consumer

The previous section indicates that the possibility that data-driven advertising practices fall under the notion of aggressive practice will depend on the image of the consumer that is the target of those practices. Concerning the latter, we can make some considerations.

3.2.1 The Average Consumer Benchmark

As seen, the UCPD adopts the test of the "average consumer" as the measure to assess the unfairness of a commercial practice. This implies that the unfairness of a practice is evaluated from the perspective of the average member of the group of consumers the practice is addressed to. This leads us to question who the average consumer is in the context of online behavioural advertising.

As suggested by Helberger, the average consumer in digital environments "must reflect in some way or other the greater technical and organisational complexity but also the changed nature of digital or digitally-enhanced products, and hence her ability to deal with that complexity".[80] In an intricate environment, where multiple businesses track, profile and target users with behavioural advertising through the use of sophisticated algorithms, the average consumer must arguably show a certain degree of awareness about the process that is taking place behind his or her screen, about the fact that the advertising is personalised according to his or her behaviour online and that potentially manipulative inferences are made on its data. If, in actuality, consumers reflect this image, is something that must be empirically proved.[81] Certainly, what seems very unlikely is that the average consumer in such a context is the one taken as a model by the UCPD itself, i.e. the consumer who is "reasonably well-informed and reasonably observant and circumspect".[82] Indeed, even before the rise of digital consumerism, the average consumer yardstick had been criticised for being a rather artificial image that does not truly reflect the actual consumer behaviour in the market and ends up failing to protect those consumers that might be more in need of protection from the law.[83] Much criticism has especially come from those who have been calling for a more "behaviourally informed approach" to the interpretation and application of the UCPD.[84] Research from behavioural economics and consumer psychology has proved that consumers

[80]The argument has been made in relation to the consumer in the IoT, but it can be reasonably extended to the consumer that is targeted by hyper-personalised advertising.

[81]We must, however, recall that in Recital 18 of the UCPD it is emphasised that the average consumer test is not a statistical test, which could be understood as to abolish the possibility to use statistical (empirical) evidence as proof, see Trzaskowski (2011).

[82]Recital 18, UCPD.

[83]Howells et al. (2018), p. 28.

[84]Tor (2013), Sibony (2014) and Duivenvoorde (2015).

show bounded rationality, will-power and self-interest and how they predictably react to certain marketing techniques and triggers (Kahneman 2011). This argument is all the more true when dealing with consumers that happen to be targeted with advertising that is customised according to their bounded rationality and will-power, and that constantly triggers desired behaviour.[85]

Besides its substantive meaning, the average consumer benchmark also shows a structural contradiction when applied to the context of OBA. As Sibony puts it, "the average consumer standard is [. . .] an inherent simplification function" and "the law [. . .] requires courts to assess whether a given practice distorts the choice pattern of the *unitary average consumer*".[86] This however is clearly in tension with the transformations that are taking place on the side of businesses.[87] The nature and the intention of online behavioural advertising, and related hyper-personalising techniques, is precisely that of breaking up with the idea of a unitary consumer, that is the product of the society of standardised consumption.

3.2.2 The Vulnerable Consumer Benchmark

Besides the general "average consumer" test, the UCPD provides for an alternative benchmark. According to Article 5(3) UCPD, wherein a commercial practice is likely only to materially distort the economic behaviour of a group of consumers which is considered vulnerable, the average consumer is modified to reflect the vulnerable consumer, provided that this is foreseeable by the trader. In such cases, the commercial practices might not be generally considered unfair in light of the average consumer's benchmark, but could still be unfair from the perspective of the vulnerable consumers.[88] OBA can therefore be evaluated according to this alternative standard, to the extent that the consumers targeted with behavioural advertising qualify as vulnerable and their vulnerability is foreseeable by the trader. Much of that will depend on the notion of vulnerability the law is willing to accept.

The UCPD covers just three types of vulnerability: (a) mental or physical infirmity; (b) age; (c) credulity. In a categorisation offered by Waddington, these vulnerabilities are described as "endogenous" since they are the result of permanent or temporary causes that are inherent to the consumer or his or her physical or mental situation.[89] This approach has been criticised for being too strict on two grounds.[90]

[85]For example, Hacker (2017), p. 271, explains how the CART test (Comprehensive Assessment of Rational Thinking), estimating specific biases and the degree of rationality of individuals, is applied in consumers profiling machine learning-based techniques.

[86]Sibony (2014), p. 77.

[87]Trzaskowski (2018).

[88]Although explicitly provided for the general clause of unfair commercial practice (Article 5), the vulnerable consumer's benchmark is also applicable to the provision on aggressive commercial practices (Article 8), see Howells et al. (2016), p. 174.

[89]Waddington (2013).

[90]Howells et al. (2018), p. 71.

First, vulnerability can be situational, that is to say, based on the circumstances in which the consumer comes to find themselves. In fact, there are many exogenous factors that are external to consumers themselves that can cause vulnerability. These are for example: lack of knowledge of the language, lack of education (in general or specific to a sector of the market) or "*the need to use new technologies with which the consumer is not familiar*".[91] Secondly, the concept of vulnerability is also relational. The literature on marketing ethics suggests that vulnerability can also be seen as a "higher susceptibility to marketing practices that create imbalances in market interactions".[92] In particular, Brenkert notes that vulnerability refers to a "combination of consumers' characteristics and the techniques used by marketers to render them particularly vulnerable";[93] Baker et al. also refer to "powerlessness due to imbalanced market interactions and consumption of marketing messages".[94]

The notion of vulnerability will also influence the second requirement, i.e. the foreseeability of the vulnerability. The vulnerable consumer benchmark only applies if the trader knows or should have known that the group targeted was going to be affected by the commercial practice. In the field of behavioural advertising, it could be reasonable—or even required under the professional diligence requirement of Article 5—to expect from the advertising companies involved in OBA a higher degree of knowledge on the capacity of data-driven advertising practices to distort the economic behaviour of consumers.

In general, we can conclude that the perspective of the UCPD on the vulnerability is relatively narrow and its elaboration somewhat limited. The eventuality that digital consumers targeted by potentially manipulative personalising advertising techniques fall under the category of "vulnerable consumers" will much rest on the elasticity of the UCPD to encompass other dimensions of the concept. A more "situational" and "relational" interpretation, for example, would arguably start a discussion on the vulnerability of consumers in the digital environment.

4 Conclusion: OBA and Unfair Manipulation Between the GDPR and the UCPD

From the analysis carried out in the previous sections, the conditions according to which online behavioural advertising is allowed are the following:

1. Under the GDPR, online behavioural advertising is lawful when data subjects *consent* to the processing of their personal data and where it is carried out for *legitimate* purposes. Additionally, OBA is *prohibited* if it significantly affects the

[91]Committee on the Internal Market and Consumer Protection (2012), p. 14.

[92]European Commission (2016b), pp. 39–82.

[93]Brenkert (1998).

[94]Baker et al. (2005).

data subject's rights and freedoms (Article 22(1) GDPR). This shall be evaluated by looking at: the intrusiveness of data collection techniques; the context's peculiarities where consumers make decisions; the advertising practice and the exploitation of consumers' personal vulnerabilities (*strong OBA*). Nonetheless, even when it has a significant effect on the data subject's rights and freedoms, OBA can be lawfully carried out when data subjects give their *explicit consent* (Article 22(2) GDPR). Finally, OBA must entail *fair* processing and data should not be processed in a way which disproportionately infringes the fundamental rights and freedoms of data subjects.

2. Under the UCPD, OBA is allowed as long as it does not, through harassment, coercion or undue influence, *impair or likely impair the average consumer's freedom of choice or conduct*, thereby causing or likely cause him to *take a transactional decision that he would not have taken otherwise* (Art. 8 UCPD). Alternatively, and secondarily, OBA can be prohibited under the general clause of unfairness, if contrary to the requirement of professional diligence, it materially distorts or likely materially distorts the economic behaviour of the average consumer it is addressed to (Art. 5).

At first glance, the most remarkable diversity between the two frameworks is based on the different perspectives adopted. The GDPR looks at the phenomenon of OBA as far as the processing of personal data is involved (*source*) and to the extent automated decision-making is carried out (*means*). On the other hand, the UCPD focuses on the external consequences of OBA specifically on consumers' decision-making (*effect*). As explained in the previous sections, data-driven advertising techniques can lead to the distortion of individual decision-making and this typically refers to the external effect of algorithmic practices on consumers. For this reason, the UCPD appears comparatively a more suitable candidate to address the problem of "market manipulation". To support this thesis, we can make additional considerations.

4.1 Consumers Data

Information about consumers is relevant for the GDPR only when it represents personal data, thus only if that information identifies or is able to identify a natural person.[95] Under the UCPD, consumer data forms part and increases the asymmetric power that characterises the trader-consumer relationship and is relevant as far as it represents actionable knowledge that traders can exploit to their advantage against consumers in order to maximise the profitability of commercial deals. The UCPD is

[95] Article 1 and 4(1) GDPR.

therefore blind, not only to the differences between personal and non-personal data but also to the distinction between data and inferences made by algorithms on data.[96] If a trader uses any information to unduly influence consumers on the condition set out by the Directive, that practice will be considered unlawful.

4.2 *Informed Consent*

Information and consent are the two central elements of contemporary information privacy and protection practices. In the GDPR, informed consent represents the prime legal basis for processing personal data. In the case of OBA, consent is generally considered as the one and only legal basis for processing and *de jure condito* seems to legitimise even the most invasive forms of automated decision-making. It is however well known that in recent years informed consent has been at the centre of a debate that highlights its inability to protect individuals' privacy against the complexity of data processing and the power of modern analytics.[97] This debate resembles the discussion among consumer lawyers about the inadequacy of information to fully protect consumers. Indeed, in consumer law too, information and consent represent the means through which the differential informative power between traders and consumers can be rebalanced.[98] The critics of informed consent have especially been influenced by behavioural sciences, which in the last decade have underlined the systematic incapacity of human beings to deal with information. More than ever now, in a setting where data controllers and traders can leverage on sophisticated methods of processing and analysis of data, individual consumers lack a similar capacity to process information. The gap is no longer and not merely informative but also technical. Although relying on the importance of information, the UCPD is nevertheless a step beyond information and consent. The Directive represents one of the two horizontal tools of consumer protection (together with the Directive on Unfair Contract Terms)[99] that provide a substantive standard of protection alternative to that of information. This brings us to consider the principle of fairness.

[96] As reported in Wachter and Mittelstadt (2019), the European Court of Justice and the Advocate General "have consistently restricted the remit of data protection law to assessing the legitimacy of the input data undergoing processing... [but not] the accuracy of decisions and decision-making processes involving personal data or to make these processes fully transparent".

[97] Koops (2014) and Mantelero (2014).

[98] Howells (2005).

[99] Council Directive 93/13/EEC of 5 April 1993 on unfair terms in consumer contracts.

4.3 Fairness

Besides its procedural dimension, the principle of fairness in the GDPR would entail a balance between data controller's interests and the fundamental rights and freedom of the data subjects, only to the extent these rights are at the core of privacy law. It is nevertheless doubtful whether this balance can also take into consideration a "right not to be manipulated". The latter remains, for now, a mere economic interest of consumers. That is why the concept of fairness implied in the UCPD seems to better fit. Here, the principle of fairness enlarges the perspective and assesses the relationship between data controllers and data subjects in its commercial dimension. In that, fairness operates as a counterbalance to the unequal negotiating power of the two parties and increases the accountability of data controllers beyond GDPR's duties.[100]

4.4 Individual and Collective Dimension

The broader perspective that characterises the fairness principle in the UCPD is also motivated by the fact that the Directive does not consider individual rights of consumers, but their collective economic interests. This stems from the indirect public objective of the Directive: unfair commercial conducts that harm consumers' economic interests indirectly harm competitors, thereby creating fragmentation in the market. Although acknowledging its societal dimension, the GDPR instead safeguards the right to data protection largely as an individual right. It has however been noted that the use of personal information and big data analytics to support decision-making exceeds the individual dimension of privacy rights and assumes a collective dimension with potentially harmful consequences for some groups.[101] This is even more evident when one considers the problem of manipulation through OBA. When a consumer is targeted with manipulative advertising and pushed into a transaction that otherwise he or she would not have made, he or she might not even notice that (after all, manipulation is all playing upon consumers by artful means). Even when he or she notices that he or she might not know it is illegal and when she knows it is illegal, he or she might fail to undertake any action, since the cost of acting will often be much higher than the cost of ignoring manipulation. Hence, it might be that for every and each individual consumer the economic harm stemming from manipulative advertising is small. However, on the societal level, the most basic principles the EU economic constitution is built upon (individual autonomy, fair competition) are being disregarded.[102]

[100]Helberger et al. (2017).

[101]Mantelero (2016).

[102]Jabłonowska et al. (2018).

4.5 Decision-Making

The conflict between the individual and collective dimension also concerns the concept of automated decision-making taken on by the GDPR. The rationale behind Article 22 is that of protecting human dignity, preventing that relevant decisions concerning human beings are made only on account of their "data shadows".[103] The prohibition of the GDPR is clearly aimed at protecting the dignity of the individual who can be potentially affected by and because of a decision taken with solely automated means (the heading itself of Article 22 refers to "automated *individual* decision making, including profiling"). On the other hand, the "decision" underlying a business' course of conduct has the potential to affect not simply one individual's right, but the interests of a multitude of subjects. This is also echoed by the mechanism of enforcement of the Directive, which traditionally has been primarily a matter of public bodies and other organisations representative of consumers' collective interest.

4.6 Final Conclusions

For all of these reasons, the UCPD could sound promising. Despite being drafted for commercial practices that took place in the pre-digital era, its technology-neutral spirit and open-textured architecture open up to novel and future-proof interpretations. One must nevertheless admit that, not only would much more guidance on the Directive be needed to cope with the new features and power relationships of digital markets (concepts such as undue influence, harassment, distortion of consumers' economic behaviours should be further elaborated and given new meanings), but also a reconsideration of consumer protection policy, that is deeply rooted in the Directive. In particular, efforts should be made to rethink the meaning and the construction of the standard of the average consumer that is tumbling before the insights coming from behavioural economics and personalised commercial practices.

References

Article 29 Working Party (2010a) Opinion 2/2010 on online behavioural advertising (WP 171) 22 June 2010

Article 29 Working Party (2010b) Opinion 4/2010 on the European Code of Conduct of FEDMA for the use of personal data in direct marketing (WP 174) 13 July 2010

Article 29 Working Party (2011) Opinion 15/2011 on the definition of consent (WP 187) 13 July 2011

[103]Mendoza and Bygrave (2017).

Article 29 Working Party (2013) Opinion 03/2013 on purpose limitation (WP 203) 2 Apr 2013
Article 29 Working Party (2014) Opinion 06/2014 on the notion of legitimate interests of the data controller under article 7 of Directive 95/46/EC (WP 217) 9 Apr 2014
Article 29 Working Party (2017) Guidelines on automated individual decision-making and profiling for the purposes of Regulation 2016/679 (WP 251) 3 Oct 2017
Baker SM et al (2005) Building understanding of the domain of consumer vulnerability. J Macromark 25:128
Bar-Gill O (2013) Seduction by contract: law, economics, and psychology in consumer markets. Oxford University Press, Oxford
Brenkert G (1998) Marketing and the vulnerable. Ruffin Series Soc Bus Ethics 1:7
Bright LF, Daugherty T (2012) Does customization impact advertising effectiveness? An exploratory study of consumer perceptions of advertising in customized online environments. J Mark Commun 18:19
Burr C, Cristianini N (2019) Can machines read our minds? Minds Mach 29:461
Burr C et al (2018) An analysis of the interaction between intelligent software agents and human users. Minds Mach 28:735
Bygrave LA (2002) Data protection law: approaching its rationale, logic, and limits. Kluwer Law International, London
Calo MR (2013) Digital market manipulation. George Wash Law Rev 82:995
Carballo-Calero PF (2016) Aggressive commercial practices in the case law of EU member states. J Eur Consum Market Law 5:255
Clifford D (2017) Citizen-consumers in a personalised galaxy: emotion influenced decision-making, a true path to the dark side? SSRN Electronic J. https://www.ssrn.com/abstract=3037425. Accessed 29 Nov 2019
Clifford D, Ausloos J (2018) Data protection and the role of fairness. Yearb Eur Law 37:130
Committee on the Internal Market and Consumer Protection (2012) Report on a strategy for strengthening the rights of vulnerable consumers, Explanatory Statement
Contissa G et al (2018) CLAUDETTE meets GDPR - Automating the evaluation of privacy policies using artificial intelligence. https://www.beuc.eu/publications/beuc-x-2018-066_claudette_meets_gdpr_report.pdf. Accessed 17 Apr 2020
Duivenvoorde BB (2015) Consumer benchmarks in the unfair commercial practices directive. Springer, Heidelberg
European Commission (2016a) Commission Staff Working Document Guidance on the implementation/application of Directive 2005/29/EC on unfair commercial practices accompanying the Communication from the Commission to the European Parliament, the Council, the European Economic and Social Committee and the Committee of the Regions – a comprehensive approach to stimulating cross- border e-Commerce for Europe's citizens and businesses, SWD/2016/0163 Final. https://eur-lex.europa.eu/legal-content/EN/TXT/?uri=CELEX%3A52016SC0163. Accessed 17 Apr 2020
European Commission (2016b) Consumer vulnerability across key markets in the European Union (Final Report). https://ec.europa.eu/info/sites/info/files/consumers-approved-report_en.pdf. Accessed 17 Apr 2020
Graef I et al (2018) Fairness and enforcement: bridging competition, data protection, and consumer law. Int Data Privacy Law 8:200
Hacker P (2017) Personal data, exploitative contracts, and algorithmic fairness: autonomous vehicles meet the Internet of Things. Int Data Privacy Law 7:266
Hanson D, Kysar DA (1999) Taking behavioralism seriously: the problem of market manipulation. N Y Univ Law Rev 74:630
Helberger N (2016) Profiling and targeting consumers in the Internet of Things – a new challenge for consumer law. In: Schulze R, Staudenmayer D (eds) Digital revolution: challenges for contract law in practice. Nomos, Baden-Baden
Helberger N et al (2017) The perfect match? A closer look at the relationship between EU consumer law and data protection law. Common Market Law Rev 54:1427

Howells G (2005) The potential and limits of consumer empowerment by information. J Law Soc 32:349
Howells G et al (2016) European fair trading law: the unfair commercial practices directive. Routledge, New York
Howells G et al (2018) Rethinking EU consumer law. Routledge, New York
ICO (2017) Feedback request – profiling and automated decision-making. April 2017. https://ico.org.uk/media/2013894/ico-feedback-request-profiling-and-automated-decision-making.pdf. Accessed 13 June 2019
Jabłonowska A et al (2018) Consumer law and artificial intelligence: challenges to the EU consumer law and policy stemming from the business. Use of artificial intelligence - Final Report of the ARTSY Project. https://artsy.eui.eu/wp-content/uploads/sites/37/2018/07/ARTSY-Final-Report.pdf. Accessed 17 Apr 2020
Kaptein M et al (2015) Personalizing persuasive technologies: explicit and implicit personalization using persuasion profiles. Int J Hum Comp Stud 77:38
Kahneman D (2011) Thinking, fast and slow. Macmillan
Koops B-J (2014) The trouble with European data protection law. Int Data Privacy Law 4:250
Law Commission (2004) Simplifying consumer legislation – A response from the Law Commission to the DTI's Consultative Document on consumer strategy. October 2004
Mantelero A (2014) The future of consumer data protection in the E.U. Re-thinking the "Notice and Consent" paradigm in the new era of predictive analytics. Comp Law Secur Rev 30:643
Mantelero A (2016) Personal data for decisional purposes in the age of analytics: from an individual to a collective dimension of data protection. Comp Law Secur Rev 32:238
Matz SC, Netzer O (2017) Using big data as a window into consumers' psychology. Curr Opin Behav Sci 18:7
Matz SC et al (2016) Money buys happiness when spending fits our personality. Psychol Sci 27:715
Matz SC et al (2017) Psychological targeting as an effective approach to digital mass persuasion. Proc Natl Acad Sci 114:12714
McStay A (2016) Digital advertising. Palgrave, London
Mendoza I, Bygrave LA (2017) The right not to be subject to automated decisions based on profiling. In: Synodinou TE et al (eds) EU internet law: regulation and enforcement. Springer, Berlin
Micklitz H-W (2018) Consumer law in the digital Eeonomy. In: Kono T et al (eds) Transnational commercial and consumer law. Springer, Singapore
Mik E (2016) The erosion of autonomy in online consumer transactions. Law, Innov Technol 8:1
Nissenbaum HF (2010) Privacy in context: technology, policy, and the integrity of social life. Stanford Law Books, Stanford
Noto La Diega G (2018) Data as digital assets. The case of targeted advertising. In: Bakhoum M et al (eds) Personal data in competition, consumer protection and intellectual property law: towards a holistic approach? Springer, Berlin
Pariser E (2011) The filter bubble: what the internet is hiding from you. Penguin Press, New York
Rhoen M (2016) Beyond consent: improving data protection through consumer protection law. Internet Policy Rev 5(1):1–15
Richards JI, Curran CM (2002) Oracles on "advertising": searching for a definition. J Advert 31:63
Sax M et al (2018) Health as a means towards profitable ends: mHealth apps, user autonomy, and unfair commercial practices. J Consum Policy 41:103
Schmechel P (2017) Verbraucherdatenschutzrecht in der EU-Datenschutz-Grundverordnung. In: Joost G et al (eds) Verbraucherrecht 2.0 - Verbraucher in der digitalen Welt. Nomos, Baden-Baden
Schwartz PM (2000) Internet privacy and the state. Conn Law Rev 32:815
Sibony A-L (2014) Can EU consumer law benefit from behavioural insights? An analysis of the unfair practices directive. Eur Rev Priv Law 22:901
Souiden N et al (2017) Consumer attitudes toward online advertising: the moderating role of personality. J Promot Manag 23:207

Stuyck J et al (2006) Confidence through fairness? The new directive on unfair business-to-consumer commercial practices in the internal market. Common Market Law Rev 43:107

Susser D (2019) Invisible influence: artificial intelligence and the ethics of adaptive choice architectures. In: Proceedings of the 2019 AAAI/ACM Conference on AI, Ethics, and Society - AIES '19. ACM Press, New York. http://dl.acm.org/citation.cfm?doid=3306618.3314286. Accessed 29 Nov 2019

Thaler RH, Sunstein CR (2003) Libertarian paternalism. Am Econ Rev 93:175

Thaler RH, Sunstein CR (2009) Nudge: improving decisions about health, wealth, and happiness. Penguin Books, New York

Tor A (2013) Some challenges facing a behaviorally-informed approach to the directive on unfair commercial practices. In: Tóth T (ed) Unfair commercial practices: the long road to harmonized law enforcement. Pázmány Press, Budapest

Trzaskowski J (2011) Behavioural economics, neuroscience, and the unfair commercial practises directive. J Consum Policy 34:377

Trzaskowski J (2018) Behavioural innovations in marketing law. In: Micklitz H-W et al (eds) Research methods in consumer law. Edward Elgar Publishing, Cheltenham

Veale M, Edwards L (2018) Clarity, surprises, and further questions in the Article 29 Working Party draft guidance on automated decision-making and profiling. Comp Law Secur Rev 34:398

Wachter S, Mittelstadt B (2019) A right to reasonable inferences: re-thinking data protection law in the age of big data and AI. Columb Bus Law Rev:494

Waddington L (2013) Vulnerable and confused: the protection of "vulnerable" consumers under EU law. Eur Law Rev:24

Wang J et al (2017) Display advertising with Real-Time Bidding (RTB) and behavioural targeting. arXiv:1610.03013. http://arxiv.org/abs/1610.03013. Accessed 29 Nov 2019

Weinmann M et al (2016) Digital nudging. Bus Inf Syst Eng 58:433

Yeung K (2017) "Hypernudge": big data as a mode of regulation by design. Inf Commun Soc 20:118

Zarsky TZ (2003) "Mine your own business!": making the case for the implications of the data mining of personal information in the forum of public opinion. Yale J Law Technol 5:1–57

Zuiderveen Borgesius FJ (2015) Personal data processing for behavioural targeting: which legal basis? Int Data Privacy Law 5:163

Zuiderveen Borgesius FJ (2016) Singling out people without knowing their names – behavioural targeting, pseudonymous data, and the new data protection regulation. Comp Law Secur Rev 32:256

Protecting Deep Learning: Could the New EU-Trade Secrets Directive Be an Option for the Legal Protection of Artificial Neural Networks?

Jasper Siems

Contents

J. Siems (✉)
Faculty of Law, University of Osnabrück, Osnabrück, Germany
e-mail: jsiems@uni-osnabrueck.de

M. Ebers, M. Cantero Gamito (eds.), *Algorithmic Governance and Governance of Algorithms*, Data Science, Machine Intelligence, and Law 1,
https://doi.org/10.1007/978-3-030-50559-2_7

Abstract Deep learning based on artificial neural networks is currently the most promising machine learning method in the field of AI. This paper distinguishes four legal protection objects of artificial neural networks per se: the training data, the topology, the weights as an expression of the trained network, and the specific training method. Both archetypical intellectual property (IP) rights, copyright and patent law, fall to some extent short of protecting these objects. This article examines whether and to what extent trade secret protection could be a suitable or supplementary legal protection tool. Trade secret protection is, among other advantages, flexible. Its greatest weakness, however, is that it allows for reverse engineering which in turn limits its application as a legal protection tool. In the case of an adaptation, trade secret law could at least temporarily supplement patent law and partially replace the classical anthropocentric copyright law in the field of deep learning.

1 Introduction

The regulation of artificial intelligence (AI) is often connected to feelings of fear: it is supposed to prevent job losses, counteract discrimination, or overcome privacy issues. At the same time, AI holds huge potential which some economists call as revolutionary as the invention of electricity.[1] The European Union's (EU) approach to counteract the complex challenges associated with the regulation of AI is to establish a more ethical kind. However, the development of any European AI industry, leave alone an ethical one, will depend mainly on private investments. If this technology is not given a fair and adequate legal protection the returns to these investments will be endangered.[2] The resulting uncertainty would almost certainly dampen investments and obstruct the industry as a whole.

Information technology (IT) develops at a rapid pace. In the context of AI and intellectual property (IP) law, there are predominantly two issues at stake: the protection of the AI technology itself and the IP-protection of by-products of AI. This paper will focus on the first aspect. It concentrates on the most promising AI technology that is currently available: deep learning (DL) performed by artificial neural networks (ANN). From a legal point of view, this article will examine the usefulness and suitability of trade secrets which constitute the intellectual stepchild of IP law. Although trade secrets gained prominence through their application in

[1]Goldfarb et al. (2018).

[2]Wenskay (1990).

protecting the core value of the intellectual property of Coca Cola as one of the largest companies of the twentieth century, as well as that of Alphabet Inc. as perhaps the most valuable company of the twenty-first century, trade secrets as protection tool are still neglected in the context of deep learning.

This paper aims to assess whether the disregard of trade secrets is justified. Firstly, it will present the technical and economic framework of deep learning that is relevant to intellectual property. Subsequently, the protection of the four objects of protection through classic IP rights, copyright and patent law will be evaluated. In the main part, the article tackles the relevance of the new EU Trade Secrets Directive 2016/943 (TSD) with regards to specific the four subject matters. It includes the advantages and disadvantages of protection through trade secrets. Lastly, a conclusion will be drawn by summarizing the results of this examination and giving an outlook for future regulation approaches in the field of trade secret law and deep learning.

2 Technical Background: A Short Overview

2.1 AI, Machine Learning and Deep Learning

The buzzwords artificial intelligence (AI), machine learning, and artificial neural networks are often used synonymously. In the understanding of this paper, they exist in a graded relation to each other. The term AI[3] describes multiform mathematical models processing methods in computer science. They are considered computer activities in a human domain which, according to anthropomorphic understanding, presuppose intelligence. Machine learning is currently an essential field in the context of AI.[4] It has a variety of methods, such as k nearest neighbour, support vector machine or the random forest method.[5] ANN, the core technology of deep learning applications, is only one of them.

[3]The EU High Level Expert Group on artificial intelligence (AI HLEG) defines AI as follows: “Artificial intelligence (AI) refers to systems that display intelligent behaviour by analysing their environment and taking actions – with some degree of autonomy – to achieve specific goals. AI-based systems can be purely software-based, acting in the virtual world (e.g. voice assistants, image analysis software, search engines, speech and face recognition systems) or AI can be embedded in hardware devices (e.g. advanced robots, autonomous cars, drones or Internet of Things applications)”; AI HLEG (2019).

[4]Ng and Soo (2017).

[5]Ng and Soo (2017).

2.2 *Deep Learning and Neural Networks: A Short Technical Overview*

Deep learning created the rehabilitation of AI in the late 2000s and early 2010s, following its breakthrough in the 1960s and subsequent downfall of popularity.[6] Deep learning is based on a learning data representation that replaced the former practise of programming if/then algorithms. The architecture and core element of deep learning applications are ANN. An artificial neural network is a biologically inspired network consisting of interconnected computing nodes (neurons) which are reacting to each other as well as external impulses.[7] Neurons are organized in different layers.[8] Neural networks consist of an input layer, several hidden layers, and an output layer.[9] The number of neurons per layer can vary and depend on the special task the ANN shall perform. Each connection between neurons is weighted individually. Moreover, every neuron has a specific activation threshold value. For a neuron to perform a particular step or transmit a signal, it must be addressed with a specific weight above its activation threshold.[10] The training data and the selected learning algorithms influence the value of the weights which represent the strength of the connections between the individual neurons.[11] The optimized set value of the weights is the most important factor for the accuracy of an ANN. It influences the ability of ANNs to predict specific outcomes, for example recognizing a stop sign in the course of autonomous driving.[12] In order to do so, the ANN needs to optimize its value of weights via training. The most common approach to train a neural network, called supervised learning, feeds the ANN with labelled data. In order to do so, one needs a sufficient amount of training data which must be further adapted for processing by the ANN. Before training a neural network, a developer must determine the so-called "topology" that constitutes the architecture of the network. This development step includes the determination of e.g. the number of hidden layers, the number of neurons per layer, and the learning and activation algorithms.[13]

The neural network is trained by repeatedly running through a database while adjusting weights.[14] The weights embody the different strengths of connections between the neurons of one layer to a neuron of the following layer.[15] The optimized

[6]The Economist (2016).

[7]Watkin and Rau (1996).

[8]Rathi (2018).

[9]Watkin and Rau (1996).

[10]Linardatos (2019).

[11]Linardatos (2019).

[12]Goldfarb et al. (2018).

[13]Prinz and Hartmann (2018).

[14]Fink (2017).

[15]Ebers (2017).

weights are the core result of the training.[16] Although considerable effort and investment are required to obtain these weights, the weighting information simply represents the amount of data that was given into the process.[17] This information can be isolated and stored.[18] Afterwards, it can be copied as many times as desired. From a technical point of view, this means that weights can either be stored separately from ANN in an external file, or implemented directly as code.[19]

3 Subject Matters and Business Models of Deep Learning

Training data is essential for the process of deep learning, especially the training of the ANN.[20] Usually, high investments flow into the acquisition of data. Furthermore, it is one of the decisive factors for prediction accuracy, thus the quality of the ANN.[21] Another central aspect is the topology of an ANN which is complex to determine.[22] The development of a topology requires a deep understanding of the matter to define the correct architecture of an ANN.[23] The principle of trial and error plays a significant role when determining a topology of an ANN.[24] The finally trained network is essentially no more than a set of weight information.[25] Despite requiring much effort, this weighted information in itself is easy to copy and not very revealing.[26] It can be inserted into an appropriate topology and immediately represent a functioning ANN even if the topology does not correspond entirely to the original training topology.[27] The training method, on the other hand, is more difficult to grasp as an object of protection.[28] There are several different approaches to training a neural network. Particularly, the choice and weight of the input categories are of decisive significance.[29] They determine how and with what relevance the network processes individual data points and thus have a considerable influence on the result.

[16]Ehinger and Stiemerling (2018).
[17]Ehinger and Stiemerling (2018).
[18]Drexl et al. (2019).
[19]Ehinger and Stiemerling (2018).
[20]Drexl et al. (2019).
[21]Drexl et al. (2019).
[22]Ehinger and Stiemerling (2018).
[23]Drexl et al. (2019).
[24]Drexl et al. (2019).
[25]Prinz and Hartmann (2018).
[26]Drexl et al. (2019).
[27]Prinz and Hartmann (2018).
[28]Lederer (2019).
[29]Rathi (2018).

Based on the brief overview of the technological background, there are primarily four objects of protection for the ANN per se: the training data set, the topology, the trained ANN and the specific training method in the broader sense.[30] These four objects of protection will function as both basis and yardstick to compare the suitability of copyright law, patent law, and trade secret law with regards to the protection of the neural network.

When considering the individual objects of protection, it is also worth taking into consideration the business model for deep learning. Currently, two different manifestations of deep learning dominate as a business model. On the one hand, ANN are offered as a service via cloud. On the other hand, they are sometimes implemented directly on digital devices as part of various software solutions.[31] The implemented solutions differ further with regards to whether the ANN is just an in-house solution or is offered to end-users. Particularly in the field of mechanical engineering, operational processes often use AI and ANN technology only in-house to improve its effectiveness and accuracy.[32]

4 Copyright Protection and ANN: A Brief Analysis

4.1 General Concept and Requirements of Copyright Law

Copyright law protects the expression of work but not the idea behind it. This phenomenon is called the *idea-expression dichotomy*.[33] Conventional software is first and foremost protected by copyright. Art. 1(1) Directive on The Legal Protection of Computer Programs - Council Directive 91/250/EEC (Computer Program Directive) states that computer programs shall be protected as literary works. To meet the requirements for protection under the Computer Program Directive, software needs to fulfil two requirements. First, it has to feature a "computer program" in the sense of the Computer Program Directive, Art. 1(1) Computer Program Directive. Second, the computer program must be original in the spirit of being the author's intellectual creation, Art. 1(3). Apart from that, ancillary copyrights such as a sui generis right for data bank makers exist. They do not require a creative element but a substantial investment.[34]

[30] See Rathi (2018).
[31] Prinz and Hartmann (2018).
[32] Goldfarb et al. (2018).
[33] Watkin and Rau (1996).
[34] Leistner (2017).

4.2 *Copyright Protection Concerning Our Subject Matters*

4.2.1 Copyright Protection for the Training Data Set

The training data set cannot be protected by standard copyright law. In general, copyright protection demands an intellectual creation. The pure collection and preparation of a training data set does not meet this criterium.[35] However, sui generis protection of the database under the EU Database Directive 96/9/EC (EU Database Directive) could apply. A sui generis database right has two central characteristics. On the one hand, it must have independent elements. On the other hand, the owner must have made a substantial investment for the procurement, examination, and representation of the data. The European Court of Justice (ECJ) twice ruled on the concept of substantial investment.[36] According to the ECJ, only the investments to obtain the contents of a database are relevant for the substantial requirement whereas investments into the creation of materials should not play a role.[37] These judgments should not lead to the conclusion that the sui generis database right does not cover machine-generated data at all. The collection and collation of data from different data sources are part of the protection.[38] Finally, processing through ANN requires a refinement of the data, which should represent a substantial investment as well.

Hence, the sui generis database protection right should apply to most input data sets. However, it should be kept in mind that this is a specific feature of EU copyright law. The US legal system, for instance, does not recognize such a property right despite some initial discussions on the topic. The lack of a sui generis database right outside the EU could significantly limit the effectiveness of such protection.[39] This is particularly relevant amidst the fact that many US-American tech firms develop AI applications.

4.2.2 Copyright Protection for the Topology

The modelling of ANN-topology is an example of application-oriented programming.[40] The topology of an ANN is represented in code within a more extensive software solution.[41] It sometimes takes only a few lines of (Python) code to define the topology of an ANN.

[35] Rathi (2018).

[36] ECJ, Case C-203/02 BHB v Hill (2004, ECT I – 10497, paras 31–42); Case C-444/02 Fixtures Marketing b Oy Veikkaus AB [2004] ECR I-10365, paras 34–49).

[37] Leistner (2017).

[38] Leistner (2017).

[39] Leistner (2017).

[40] Prinz and Hartmann (2018).

[41] Prinz and Hartmann (2018).

The "computer program" criterion in Art. 1(1) Computer Program Directive is deliberately not defined in the directive. The term was left open to interpretation to facilitate rapid technological advancement. Many legal scholars define computer programs as a "set of statements or instructions to be used directly or indirectly in a computer to bring about a certain result".[42] The terms "statements," as well as "instructions" in the meaning of this definition describe a way of executing commands or interpreting individual elements through a computer.[43]

The topology indeed fulfils such an instructing function. It is the basis for the creation of a trained ANN. The developer determines the number of layers and neurons as well as the used training algorithms through source code. Therefore, this source code carries out instructions which allow the systems to process the data.[44] The processing of data are in the first step the "learning" and in the second step the "applying" of an ANN.[45]

4.2.3 Copyright Protection for the Trained ANN

The protection of the trained network based on copyright law turns out to be quite difficult on closer examination. European copyright law puts the human and his intellectual creation at the centre. This anthropocentric approach also applies explicitly to software copyright. This requirement of an intellectual creation poses enormous difficulties for copyright law in protecting the trained network.

The trained ANN embodies a collection of data that associates neurons.[46] They no longer have much in common with the source code of conventional software. Almost every traditional software is programmed explicitly which means that individual steps to solve a problem are determinate and programmers translate them into the source code. Therefore, it is crucial that programmers know in advance of the actual programming how to describe a particular problem with the help of a programming language. Through the source code, the computer receives instructions as to which arithmetic operations it has to perform when, and in which manner. The weight information of the connection between the neurons, however, is not capable of instructing a machine to carry out a specific process.[47] The weight information are merely data structures.[48] They are of crucial significance for the behavior of a computer. However, the weight information per se does not instruct the computer to do anything.[49] As a result, under the classical understanding of the term

[42]Rathi (2018).

[43]Rathi (2018).

[44]Ehinger and Stiemerling (2018).

[45]Ehinger and Stiemerling (2018).

[46]Ehinger and Stiemerling (2018).

[47]Prinz and Hartmann (2018).

[48]Ehinger and Stiemerling (2018).

[49]Ehinger and Stiemerling (2018).

"computer program" the weights fall out of the scope of Art. 1(1) Computer Program Directive.

The weights must further be regarded as an intellectual creation due to Art. 1 (3) Computer Program Directive. Therefore, a human must be the creator of the work.[50] The author can use a machine as a tool, and the result of the creation process does not have to be entirely predictable for the author.[51] When there is no human contribution and selection anymore, copyright protection cannot be granted. Some legal scholars argue that this is often the case for weights as a result of the training of a neural network.[52] Considering the increasing autonomy of ANNs, this question will most likely become more and more virulent. It is especially true for newer, less established learning techniques, such as reinforcement or unsupervised learning which lead to an increasing separation of program and programmer. When the software is writing software makes further progress, the anthropocentric software copyright law approach could finally be overstrained.[53] A current example is the technique of Generative Adversarial Networks (GAN). Generative Adversarial Networks consist of two ANNs that perform a zero-sum game.[54] One of them creates candidates (the generator) while the second neural network evaluates the candidates (the discriminator).[55] The generator's goal is to learn to generate results according to a specific distribution. The discriminator is trained to distinguish the results of the generator. The objective function of the generator is then used to produce results that the discriminator cannot distinguish. The procedure should result in a generated data distribution which is supposed to align with the real allocation of the data set or given archetype.

Finally, the copyright protection of trained neural networks holds massive enforcement troubles.[56] Copyright law always protects the specific expression of an idea.[57] It does not cover the underlying concept of the expressed idea.[58] Hence, a copyright is not infringed by someone who copies only the underlying scheme of a network.[59] Someone could replicate the exact set of weights and modify them negligibly.[60] This procedure does not necessarily result in reduced performance of the neural network. Moreover, it might not violate any copyrights.[61] Also, it could be

[50]Ehinger and Stiemerling (2018).
[51]Ehinger and Stiemerling (2018).
[52]Ehinger and Stiemerling (2018).
[53]de Wachter (2010).
[54]Rohner (2019).
[55]Rohner (2019).
[56]Rathi (2018).
[57]Wenskay (1990).
[58]Wenskay (1990).
[59]Rathi (2018).
[60]Rathi (2018).
[61]Wenskay (1990).

difficult to prove when a neural network is copied. The ANN, and especially the set of weights, lack any references to the original authorship.[62]

4.2.4 Copyright Protection for the Specific Training Method

Copyright protection is probably not applicable for training methods. Copyright law protects the expression of an idea but not the idea itself. Hence, it is difficult to find a reference point for the expression of a specific training method embodying an idea.

5 Patent Law: The Central IP Right Concerning Deep Learning?

Many scholars identify patent law with good reason as the central element for protecting innovation in the age of digitalization.[63] A patent is an examined exclusive right to technical teaching. There are three main requirements for granting a patent (patentability). First, there must be a patentable subject matter (invention in a technical field). Second, at least one aspect of the invention must be "novel". That means the invention must involve an inventive step. Finally, it must be available for industrial application.

5.1 Patent Protection for the Training Data Set

Training data is either pure data or some other information. It is merely a collection of data and does not represent technical teaching for the solution of a technical problem. Therefore, patent protection will in general not apply.

5.2 Patent Protection for the Topology

As a rule, the creation of topology is application-related and programmed by humans. However, automatic generation is already technically feasible.[64] If humans create the topology for a specific (technical) application patent protection seems likely.

---

[62]Rathi (2018).

[63]McGuire and Nikolidaki (2016).

[64]Sun et al. (2018).

5.3 *Patent Protection for the Trained Neural Network*

The trained neural network based on computational models falls per se out of the scope of European patent law. The set of weights as the expression of a trained ANN is of an abstract mathematical nature.[65] Therefore, the weights as such are not patentable. However, Art. 64(2) European Patent Convention (EPC) states that if the subject-matter of the European patent is a process, the protection conferred by the patent shall extend to the products directly obtained by such means. The weights are the direct product of the training process and thus fall into the scope of the protection. Every modification of the weights ends the protection.

Trained ANN can obtain patent protection when combined with a specific application in a technical field.[66] Here, the inventive achievement would consist for instance in solving a technical problem with an deep learning application.[67] Moreover, patent protection may apply for the neural network if it simulates a technical procedure. As an example, the European Patent Office (EPO) mentions in its guidelines the use of a neural network in a heart-monitoring apparatus for the purpose of identifying irregular heartbeats.[68] Therefore, patent law will protect the majority of neural network applications. However, a patent is not granted for non-technical trained ANN applications such as classifying text documents with respect to their textual content.[69] Patent protection applies to ANN in a large number of cases as long as they are combined with a technical application. Consequently, patent law is the central IP law for the protection of trained ANN.

5.4 *Patent Protection for the Specific Training Method*

When training an ANN, the data used is often unique. This aspect can meet the "novelty" criterion under patent law.[70] A patent can be granted when a specific training method is well represented in the patent claim. Also, data is typically technical, which means that the patent exclusion criteria stating that ANN are of

[65]European Patent Office (2018).

[66]European Patent Office (2018).

[67]Lederer (2019).

[68]European Patent Office (2018).

[69]The European Patent Office issued new guidelines regarding the patentability of AI and machine learning in the year 2018, Guidelines for Examination Part G 3.3.1. The EPO states that AI and machine learning methods "merely refer to abstract models or algorithms and thus do not, on their own, necessarily imply the use of a technical means. This has to be taken into account when examining whether the claimed subject-matter has a technical character as a whole." However, where a machine learning method service a technical purpose the EPO holds the view that patent protection could apply.

[70]Lederer (2019).

an abstract mathematical nature does not apply.[71] To what extent the data must be disclosed for the patent registration can be problematic.

6 Neural Networks and Trade Secrets

6.1 General Requirements for Protection Under the Trade Secret Directive

Until now, there has been no uniform European approach to the classification of trade secrets in the EU. In Germany, trade secrets were anchored in competition law as criminal law provisions, whereas in Sweden there was a special law for trade secrets, and in France and Italy they were regulated by IP law. This fragmentation is now changing as the Trade Secrets Directive 2016/943 harmonizes the protection of trade secrets across the EU.

Art. 2 TSD specifies three requirements for the concept of a trade secret: The information must be secret (1), of commercial value (2) and subject to reasonable steps to keep it secret (3). The directive has a somewhat unique regulatory system. It first lists lawful acts in art. 3, then unlawful acts in art. 4, and exceptions in Art. 5, with regards to the unlawful acquisition, use, and disclosure of trade secrets.

6.1.1 Secret Information

The Trade Secret Directive protects information in the broadest sense.[72] However, the regulatory purpose of the directive to promote innovation, in conjunction with recital 14, suggests that private secrets should be distinguished from trade secrets.[73] Information must be secret in the that it is not generally known or easily accessible to persons in the circles involved with the nature of the information concerned, Art. 2 (1)(b). Therefore, the Trade Secret Directive envisages a relative concept of secrecy. The practical possibility of access for a certain group of people is decisive for the assessment of secrecy.[74] If the information consists of several components it is not even obvious if some of the components are public knowledge. Although every fact per se could be public knowledge, the specific arrangement of facts could embody a secret for the public.

[71] Thiel (2019), Ohly (2019) and Lederer (2019).

[72] Ohly (2019).

[73] Ohly (2019).

[74] Ohly (2019).

6.1.2 Commercial Value

The information must have commercial, whereas a potential value should be sufficient. New technical knowledge can also have commercial value if it is still unclear whether it can be marketed.[75]

6.1.3 Reasonable Steps

Finally, the trade secret holder must take reasonable steps to protect secrecy. It is not quite clear what this means in detail, as there is no case law on this subject.[76] However, the nature and scope may depend on how valuable the secret is overall and for the company, the size of the company, and the cost and customary nature of the measures.[77] Reasonable steps could, for example, be appropriate technical and organizational measures as laid down in Art. 32 General Data Protection Regulation (GDPR).[78]

This criterion results from two considerations. On the one hand, it serves to prove the holder's interest in protection. In this way, he can and should make clear that he is interested in the protection of the information concerned.[79] Even more important is the warning function of the reasonable protective steps for third parties. In this way, third parties should be aware that they can be held liable if they unlawfully acquire a trade secret.[80]

6.1.4 The Allowance of Reverse Engineering

Art. 3 TSD regulates lawful acts and thus limits the prohibitions in Art. 4.[81] If one acquires the secret by an act mentioned in Art. 3 TSD, it is lawfully acquired. This allowance rule initially includes the independent discovery or creation of the trade secret, Art. 3(1)(a) TSD. Hence, trade secret protection is not an exclusive right. Rather, it prevents access from of misappropriated acts in the case of unfair competition.[82] The Trade Secrets Directive explicitly allows the acquisition of a trade secret through reverse engineering in Art. 3(1)(b). Although reverse engineering can be contractually prohibited by Art. 4(3)(c) TSD, a contractual exclusion does not apply to third parties. Reverse engineering describes the process of extracting the design

[75]Ohly (2019).

[76]Thiel (2019).

[77]Scheja (2018).

[78]Scheja (2018).

[79]McGuire (2019).

[80]McGuire (2019).

[81]Ohly (2019).

[82]Ohly (2019).

elements from an existing system or an industrially manufactured product by examining its structures, states, and behavioural patterns. Allowing reverse engineering could be problematic concerning ANNs. Unlike conventional software neural networks in general are comparably easy to engineer reversely.[83] This applies in particular when competitors can observe what input data enters the network and what output follows consequently.[84] The competitor may then be able to pick up the underlying topology as well as the approximate trained weights. In neural networks, a relatively small number of queries (650–4500) are sufficient to reverse engineer the models used fairly accurately and sometimes even perfectly.[85]

6.2 *Further Legal Questions Arising from Trade Secret Protection*

The secrecy could be questioned if neural networks operate in a modern cloud architecture. As a result, the information shall remain secret when stored in a cloud and shared with partner companies during a cooperation project.[86] The purpose of the directive is to foster innovation through trade secret protection. Accordingly, the (relatively) secret character of information shall remain if the extension of confidants is justified by the purpose of the business undertaken.[87] The historical background of the directive supports this legal view. Based on the Agreement on Trade-Related Aspects of Intellectual Property Rights (TRIPS), trade secret protection roots in the law against unfair competition. The EU directive was highly influenced by the TRIPS.[88]

In general, difficulties arise when it comes to the legally secure allocation of the trade secret holder.[89] The directive offers only a vague legal definition in Art. 2(2): "any natural or legal person controlling a trade secret".[90] Considering this definition, it remains unclear whether only one person can be a trade secret holder or whether this extends to licensees, participants in joint ventures or other persons to whom the information has been disclosed with consent.[91] A joint controllership of trade secrets through many parties, e.g. in the context of Big Data, shall be possible. However, a

[83] Goldfarb et al. (2018).

[84] Goldfarb et al. (2018).

[85] Tramèr et al. (2016).

[86] Sagstetter (2019b).

[87] Sagstetter (2019b).

[88] Surblyte (2017).

[89] Surblyte (2017).

[90] Sagstetter (2019a).

[91] Alpin (2017).

licensee cannot be the trade secret holder as she uses a third party right. Therefore, the licensee is restricted by its contract and does not control a trade secret.[92]

6.3 *Trade Secret Protection and Artificial Neural Networks*

6.3.1 Legal Protection per se

All objects of protection (input data, topology, trained ANN, and specific training method) should fulfil the requirements of trade secret protection. Were it possible to keep the respective object of protection secret, trade secret law would offer sufficient protection for ANNs. It is clear that the individual protection objects of an ANN are of "commercial value". All of them require, as described above, considerable financial investments. The nature and extent of the reasonable step for trade secret protection depend on the individual ANN and company. However, the company is likely to implement them regularly in the area of deep learning.

6.3.2 Limitations of Trade Secret Law Protection

Trade secret protection exists as long as the concerned information is not obvious. Generally, the main goal of the new Trade Secrets Directive is to foster innovative development before market entry.[93] Therefore, the directive also allows reverse engineering.[94] This speaks prima facie against realizing a business model primarily based on trade secret law protection. At the same time, trade secret law protection may be the only alternative for certain aspects of the ANN. In some cases, no other IP protection is applicable. Under the current legal regime it is likely that for some, especially non-technical, business models neither copyright nor sufficient patent protection applies. This lack of IP protection means that in these specific cases, the companies must account for their IP needs in their business model.

Trade secret law is primarily intended for the period preceding market entry. The goal is to ensure market access by maintaining the lead in a market. During this period, the company as a trade secret holder has optimal means of controlling access to secrets. A cloud-based solution corresponds most closely to this situation. Here, the company offering the ANN retains control over the ANN, its topology and weights. Accordingly, it is more challenging to reverse engineer a cloud-based AI service compared to one embedded in an end device. With corresponding general terms and conditions prohibiting reverse engineering, there would be a severe possibility of commercializing ANN based on the right to protect secrets. However,

[92]McGuire (2019).

[93]McGuire (2019).

[94]McGuire (2019).

this would have to be accompanied by technical protective measures. In the case of ANNs implemented in the device, one would primarily have to rely on contractual regulations since control over the ANN would be given away. Also, with an online version of an ANN, finding out weights is only partially beneficial for the competitor. The network is continuously evolving through new input and the weights are continuously changing.

A further challenge for protection primarily based on trade secret law could be the conflict with the (transpareny) requirements of the GDPR. Art 13(2)(f) GDPR, Art. 14(2)(g) GDPR and Art. 15(1)(h) GDPR all contain transparency obligations for data processing.[95] This transparency approach of the GDPR could clash with the confidential character of trade secret law. According to GDPR provisions mentioned above, the data subject depends upon meaningful information about the logic involved, as well as the significance and the envisaged consequences of such processing for the data subject. In particular, it is questionable whether the controller can be obliged to disclose the calculation formulas and parameters used in an automated decision or even the algorithms themselves.[96]

A duty to disclose the used algorithms and mechanisms does not follow from the GDPR. The purpose of transparency provisions also argues against an interpretation that includes the disclosure of trade secrets. Art. 13–15 GDPR provides “meaningful information about the logic involved”. These provisions do not, however, require that the involved logic is disclosed per se. Thus, it is a matter of describing the underlying principles of the ANN. It entails the explanation of how the ANN meets predictions, and which yardsticks it applies thereby, as well as from which premises it proceeds.[97] This approach is meant by the prerequisite of “meaningful information”. For example, the specification of the exact values of the weights as training results would not be meaningful for the person concerned. Based on these weights, he or she cannot trace whether and in what way personal data is processed.

6.4 Trade Secrets as a Legal Protection Tool Compared to Classic IP Rights

Trade secret law would first and foremost supplement patent law which could happen in several ways: First as a gap-filler, for when patent law (and copyright law) does not apply, second as an alternative when the protection of trade secret law is preferred for strategic reasons and finally, partly as a substitute for copyright law. However, trade secret law can only succeed if a well-thought-out contract clause supplements it and if the specific ANN application can be protected by trade secrets.

[95] Häuser (2018).

[96] E.g. for scoring algorithms: Rühlicke (2019).

[97] Häuser (2018).

Trade secret law shows enormous flexibility as a legal protection tool. This flexibility features two forms: legal and real-world flexibility. The Trade Secret Directive consists of many open and indeterminate legal terms. This enables an adaptive approach when protecting the different protection objects of an ANN. Trade secret law applies to all the objects of protection that have been elaborated in this paper. It is not necessary to examine on a case-by-case basis whether an ANN-based development is technical in the sense of patent law. Know-how protection is particularly crucial when programming is shifted further in preliminary stages or wholly automated. There is no need to prove a human contribution to programming as in copyright law. Notably, the Trade Secret Directive leaves more detailed questions, such as the allocation of the trade secret holder, to the parties concerned.[98] This approach seems appropriate since these parties are experts and generally know best about their needs and particular interests within the framework of their legal relationship.[99] The flexibility argument corresponds to the fact that know-how protection is more far-reaching and less definite. It already includes smaller and less significant components of ANN. In particular, the protection is independent of the manifestation of an ANN component. This argument for broadness (promotion of innovation/technical-economic progress) stands against the low level of protection (head start advantage: non-disclosure).

Compared to patent law and copyright law, trade secret law represents a lower level of protection. From a regulatory point of view, this seems appropriate for ANNs per se. Often, only small technical progress is visible, for example by optimizing specific parameters of an ANN.[100] Thus, the weaker trade secret law is better suited to the gradual development of deep learning technology than the strong patent law. Particularly in the IT sector, there is already an unmanageable thicket of patents. This development regularly leads to innovation-inhibiting procedures and disputes.

Unlike a patent whose granting period may take several years, trade secret protection occurs by default. Patent applications can quickly become expensive, especially for small and medium-sized enterprises. Trade secret protection applies without any substantial costs. Thus, it only requires appropriate secrecy measures, some of which already exist in most company. In trade secret law, it is sometimes easier to detect infringements compared to copyright law. Particularly, the trade secret is also protected if the end product is not the same, but in all cases, if the trade secret has been used in its creation.

Finally, the recourse to trade secret law prevents the establishment of a new legal regime such as data ownership. The concept of data ownership includes some kind of exclusive assignment of data.[101] When dealing with relatively new phenomena in

---

[98] Sagstetter (2019a).

[99] Sagstetter (2019a).

[100] Goldfarb et al. (2018).

[101] Fezer (2017).

the data economy, it seems reasonable to refrain from creating new laws.[102] This goes particularly for the early stages of new business models as otherwise the law could rather hinder than foster innovation.[103] Trade secret law could, in this case, act as a bridge alongside other existing legal instruments. It could close current gaps until the authorities develop a functioning and sustainable concept for IP regulation with respect to data, information, and AI.

7 Conclusions

7.1 Optimization of the Legal Regime

Although trade secret protection will remain a complementary legal protection tool, some clarification of the legal framework could enhance its effectiveness. The establishment of standard contractual clauses as known from data protection law in Art. 46 GDPR could reduce legal uncertainty.[104] Notably, the questions of trade secret holders could be tackled through a trade secret register maintained by the joint venture participants.[105] This register could, as an idea, even rely on blockchain technology.[106] Reverse engineering could be curtailed by two approaches. Since neural networks show similarities to conventional software, the same rules in the field of reverse engineering with regards to computer programs could apply to them. Therefore, reverse engineering would, except for particular use cases (among other things, interoperability), only be possible with the consent of the trade secret holder. Another approach would be to regard the simple imitation of a neural network through reverse engineering as an act of unfair competition.

7.2 Results

This brief analysis shows that trade secret law could function as an additional legal protection tool in combination with patent law. By doing so, it would to some extent substitute copyright law. This substitution would not imply a massive change of concept when protecting digital value in the software industry. The IP protection arising from copyright law is mainly related to the secrecy of the software source code. The same applies to the weights of a neural network under the Trade Secret Directive. To fill this gap, the Trade Secret Directive needs some optimization. If

[102] von Lewinski (2018).

[103] von Lewinski (2018).

[104] Sagstetter (2019a).

[105] Sagstetter (2019a).

[106] Sagstetter (2019a).

improved, the Trade Secret Directive could function as a bridge law until the regulating authorities have gained clear insight into the AI industry and the data economy. From today's point of view, this approach seems preferable compared to the introduction of new and potentially immature laws such as the concept of data ownership.

References

Alpin T (2017) Trading data in the digital economy: trade secrets perspective. In: Lohsse S et al (eds) Trading data in the digital economy: legal concepts and tools. Nomos, Baden-Baden, pp 59–74

de Wachter J (2010) Software written by software – is copyright still the appropriate tool to protect IT? CRi 11:12–20. https://doi.org/10.9785/ovs-cri-2010-12

Drexl J et al (2019) Technical aspects of artificial intelligence: an understanding from an intellectual property law perspective. Max Planck Institute for Innovation and Competition Research Paper No. 19-13, Version 1.0, October 2019, Munich. https://ssrn.com/abstract=3465577. Accessed 29 Apr 2020

Ebers M (2017) Autonomes Fahren: Produkt- und Produzentenhaftung. In: Oppermann BH, Stender-Vorwachs J (eds) Autonomes Fahren: Rechtsfolgen, Rechtsprobleme, technische Grundlagen. C.H. Beck, München, pp 93–125

Ehinger P, Stiemerling O (2018) Die urheberrechtliche Schutzfähigkeit von Künstlicher Intelligenz am Beispiel von Neuronalen Netzen. CR:761–770. https://doi.org/10.9785/cr-2018-341206

European Patent Office (2018) Guidelines for examination. https://www.epo.org/law-practice/legal-texts/html/guidelines2018/e/g_ii_3_3_1.htm. Accessed 12 Dec 2018

Fezer K-H (2017) Repräsentatives Dateneigentum. Ein zivilgesellschaftliches Bürgerrecht, Studie im Auftrag der Konrad-Adenauer-Stiftung e. V. zum Thema "Einführung eines besonderen Rechts an Daten". https://www.kas.de/de/einzeltitel/-/content/repraesentatives-dateneigentum. Accessed 29 Apr 2020

Fink L (2017) Big data and artificial intelligence. ZGE:288–298. https://doi.org/10.1628/186723717X15069451170874

Goldfarb A et al (2018) Prediction machines: the simple economics of artificial intelligence. Harvard Business Review Press, Cambridge

Häuser M (2018) DSGVO und KI: Gefahr für Geschäftsgeheimnisse? https://www.lto.de/recht/kanzleien-unternehmen/k/dsgvo-ki-geschaeftsgeheimnisse-unternehmen-schutz-offenlegung/. Accessed 26 June 2019

High Level Expert Group on AI (AI HLEG) (2019) A definition of AI: main capabilities and disciplines. Brussels, 9 Apr 2019. https://ec.europa.eu/newsroom/dae/document.cfm?doc_id=56341. Accessed 29 Apr 2020

Lederer TL (2019) Patentierung im Bereich Künstliche Intelligenz. GRUR-Prax:152–154

Leistner M (2017) Big Data and the EU Database Directive 96/9/EC: current law and potential for reform. In: Lohsse S et al (eds) Trading data in the digital economy: legal concepts and tools. Nomos, Baden-Baden, pp 25–58

Linardatos D (2019) Künstliche Intelligenz und Verantwortung. ZIP:504–509

McGuire M-R (2019) Neue Anforderungen an Geheimhaltungsvereinbarungen? WRP:679–688

McGuire M-R, Nikolidaki S (2016) Der Schutz technischer Innovationen im Binnenmarkt. IPRB:152–155

Ng A, Soo K (2017) Numsense! Data science for the layman: no math added. Annalyn Ng and Kenneth Soo, Singapore

Ohly A (2019) Das neue Geschäftsgeheimnisgesetz im Überblick. GRUR:441–451

Prinz M, Hartmann F (2018) Immaterialgüterrechtlicher Schutz von Systemen Künstlicher Intelligenz. WRP:1431–1438

Rathi S (2018) Revival of the neural networks and the intellectual property nightmare: workshop proceedings of the 14th International Conference on Intelligent Environments. In: Chatzigiannakis I et al (eds) Intelligent environments 2018: workshop proceedings of the 14th International Conference on Intelligent Environments. IOS Press Incorporated, Amsterdam, pp 275–284

Rohner T (2019) Der Schutz von KI-Schöpfungen im schweizerischen Urheberrecht. ZGE:33–85. https://doi.org/10.1628/zge-2019-0003

Rühlicke L (2019) Die Geheimhaltung von Scoring Algorithmen. In: Maute L, Mackenrodt M-O (eds) Recht als Infrastruktur für Innovation. Nomos, Baden-Baden

Sagstetter T (2019a) Big Data und der europäische Rechtsrahmen: Status quo und Reformbedarf im Lichte der Trade-Secrets-Richtlinie 2016/943/EU. In: Maute L, Mackenrodt M-O (eds) Recht als Infrastruktur für Innovation. Nomos, Baden-Baden

Sagstetter T (2019b) Digitaler Strukturwandel und Privatrecht - Neue Regeln für Big Data und Co.? In: Husemann T et al (eds) Strukturwandel und Privatrecht: Jahrbuch Junge Zivilrechtswissenschaft. Nomos; Helbing & Lichtenhahn, Baden-Baden

Scheja K (2018) Schutz von Algorithmen in Big Data Anwendungen. CR:485–492

Sun Y et al (2018) Automatically designing CNN architectures using genetic algorithm for image classification. https://arxiv.org/abs/1808.03818v1. Accessed 29 Apr 2020

Surblyte G (2017) Data-driven economy and artificial intelligence: emerging competition law issues? WuW:120–127

The Economist (2016) (Special report) From not working to neural networking: the artificial boom is based on an old idea, but with a modern twist. https://www.economist.com/special-report/2016/06/23/from-not-working-to-neural-networking. Accessed 17 Apr 2020

Thiel L (2019) Das neue Geschäftsgeheimnisgesetz (GeschGehG) – Risiken und Chancen für Geschäftsgeheimnisinhaber. WRP:700–703

Tramèr F et al (2016) Stealing machine learning models via prediction APIs. https://arxiv.org/abs/1609.02943v2. Accessed 29 Apr 2020

von Lewinski K (2018) Regulierungsbedarf und Regulierungsfelder von algorithmischen Systemen. InTer:168–176

Watkin T, Rau A (1996) Intellectual property in artificial neural networks - in particular under the European Patent Convention. ICC:447–469

Wenskay DL (1990) Intellectual property protection for neural networks. Neural Netw 3:229–236. https://doi.org/10.1016/0893-6080(90)90092-Y

Chinese Copyright Law and Computer-Generated Works in the Era of Artificial Intelligence

Ying Ye and Mike Adcock

Contents

Abstract Artificial intelligence (AI) technology has been widely applied by innovative industries with a large number of outputs indistinguishable from human works. These outputs raise debates on the legal status of computer-generated works (CGWs) under copyright law, questions which are directly related to certain key issues of copyright law, including copyrightability and authorship. This article, based on the analysis of China's copyright system, raises the question about whether the judgment criterion of the originality of computer-generated works should take into account the author's personality and creativity. It argues that some existing specific rules of Chinese copyright law, such as "the work created in the course of employment" and "the work of legal person or entity", allow for the separation of the authorship and ownership. Under the basic principle of "the author is the owner of copyright", these rules value the importance of investors, which provides a possible approach to the protection of CGWs by the current copyright law in the future.

Y. Ye (✉) · M. Adcock
Durham Law School, Durham, UK
e-mail: ying.ye@durham.ac.uk; mike.adcock@durham.ac.uk

M. Ebers, M. Cantero Gamito (eds.), *Algorithmic Governance and Governance of Algorithms*, Data Science, Machine Intelligence, and Law 1,
https://doi.org/10.1007/978-3-030-50559-2_8

1 Introduction

This section illustrates that the technology of artificial intelligence (AI) has been widely applied in our current society to produce creative works. Human involvement varies between AI-generated works and works that are created with the assistance of computers, which has led to these works having different legal definitions. In turn, this has led to new copyright concerns.

The use of AI is widespread through different sections of our society, resulting in various types of social productions. On the internet, it helps commercial actors to select contents to cater to the preference and demand of consumers, disseminate targeted advertisements and predict users' behaviour to achieve better trading results. AI is also involved in the field of creation/creative works. With this advanced technology, a computer can learn to create a new "content", especially in traditional artistic and literary forms. One of the most notable examples is a project led by a Dutch team, named "the next Rembrandt", whose aim is to use AI technology to create a new "Rembrandt's work" in the twenty-first century. This project, with the help of a facial recognition system, resulted in a new masterpiece in the style of Rembrandt, which would have likely been an impossible task to achieve previously. In addition to the creation of art, media companies are using AI to generate news,[1] while music brands provide rapid and short music-producing services utilising artificial intelligence. In China, this technology has been applied to many areas, including the creation of the Xiaoice application, an active online "poetry writer" that automatically generates modern Chinese poems with designated images selected by the users. Another AI design software, Lu Ban, provided by Alibaba, offers users a variety of poster-designing services. These instances illustrate that AI technology has widely utilised in business areas related to human creations, and an ever-increasing number of companies in the creative field are trying to generate new contents in new commercial areas, utilising this technology.

This trend has raised new copyright concerns regarding works generated by or with the help of AI technology. Unlike works produced with the assistance of computers, works generated by computers may challenge the traditional copyright legal system. In previous times, computers were usually regarded as the tool which plays the role of an "assistant" for human authors.[2] These kind of works, created by human authors with the assistance of computers, are termed computer-assisted works (CAWs). However, AI-generated works, such as the ones produced by Xiaoice or Lu Ban, belong to a new category, named computer-generated works (CGWs). These are works autonomously generated by computers without the creative participation of human authors. In the process of creating CGWs, computers

[1]Companies like Associated Press, Forbes, New York Times, Los Angeles Times, and Tencent.

[2]For example, in the word processing software, the identity of the author is clear that the authorship is allocated to the natural person who actually organises the content and provide expressions, not the computer.

apply their "learning ability", such as machine learning and deep learning. In addition, computers can also extract information from large amounts of data and apply this to the generation of content.[3] The result is that human beings put less and less intellectual intervention into the creative process, which becomes more dependent on AI. A more automated content production process is a vital difference between CAWs and CGWs. This raises fundamental issues in copyright, namely copyrightability and authorship. Only a few countries have direct regulations governing these issues. UK copyright law, for example, provides a legal definition of CGWs in the CDPA Act 1988 stating ""computer-generated", in relation to a work, means that the work is generated by the computer in circumstance such that there is no human author of the work."[4] The Act also provides regulations about the definition of authorship of CGWs while excluding moral rights from these works.[5] However, legislation in most countries and regions has not yet recognised copyright in this sort of work.[6] This prevents CGWs from being a conventional type of work in a copyright sense under many jurisdictions,[7] which means copyright protection for CGWs is uncertain in many states.

China also faces the same issues. The reason why Chinese society has started to pay attention to copyright issues regarding CGWs is the ever-increasing number of CGWs appearing on the market, many of which are hard to distinguish from human works.[8] This phenomenon has led to society discussing the legal status of these works. However, there is no direct legal provision governing CGWs in Chinese copyright law, which has led to increased discussion related to questions such as: can or should the copyright law protect CGWs? Are CGWs original? Who is the author of CGWs? With the expanding use of AI and the growing number of AI-generated works, these questions require answers from the aspect of copyright law.

In conclusion, AI technology is being increasingly applied to the creative fields, as technology improves the efficiency and effectiveness of content productions. This has resulted in some creations, which used to require human intellectual intervention, now be created as CGWs. Therefore, the copyrightability of CGWs has become a major issue, and many copyright-related questions regarding CGWs and AI technology need to be discussed and resolved.

In order to illustrate this, the article will first provide an overview of the development of AI and its related copyright issues, followed by an overview of

[3]The next Rembrandt program is the typical manifestation of application of "learning ability". In this project, experts fed a big amount of data of Rembrandt's works to the computer, which generated outputs of Rembrandt' style (artistic patterns) with artificial intelligence.

[4]Section 178, CDPA 1988.

[5]Section 9(1), Section 79 and Section 81 CDPA 1988.

[6]Including the EU and US, See: Case C-145/10 *Eva-Maria Painer v Standard Verlags GmbH & Others* [2010] ECR I-12533; U.S. Copyright Office, Compendium of U.S. Copyright Office Practices § 313.2 (3d ed. 2017).

[7]For instance, even the Berne Convention does not involve computer-generated works. See: Article 2, Berne Convention.

[8]Tecent Research Institute (2017).

different views among Chinese scholars and judges on copyright protection of AI-generated works. The article will then separately discuss the issues of copyrightability and authorship. Finally, a conclusion will be reached.

2 Different Views on Copyright Protection of CGWs in China

It is still a controversial issue among Chinese scholars and judges whether CGWs can or should be granted copyright protection. One of the mainstream views is in favour of copyright protection of CGWs. According to Xiong (2017) and Yi (2017), whether CGWs are original in the copyright context requires two points, namely, the minimum degree of creativity and the nature of the generation process. Xiong believes that since human works and CGWs are indistinguishable, it shows CGWs can achieve the minimum degree of creativity. From the perspective of industrial development, Cao (2016) believes that if CGWs are not considered as works, a large number of CGWs will enter the public domain and investment behaviour will be affected, which will not be conducive to the entire industry.

However, there are also many scholars who object to this view. Wang (2017) believes that, as CGWs and works of human authors are hard to distinguish, the point of copyrightability should be focused on the process of producing the work. The works are outputs of the application of computer algorithms, which cannot embody the creator's spirit and personality. Therefore, computer-generated works should not be protected by copyright law. This opinion is similar to that of the Beijing Internet Court in the *Beijing Film Law Firm v. Baidu* case in 2019.[9] In the first-ever case of the copyright protection of AI-generated works in China, the court held that the graphs claimed by the law firm are not graphic works in a copyright sense, because the differences shown by the graphics "are derived from differences of data rather than creation".[10] The court pointed out that, if different users choose the same data, they will obtain the same results by using the same software. Therefore, the graphics in this case lack originality and should not be given copyright protection. As for the text automatically generated by computers, the court held that "creation and completion by natural persons should still be a necessary condition for "works" defined in the Copyright Law".[11] So even though the contents generated by computers reflect a certain level of originality, they are not "works" because of the lack of human authors in the process of creation. However, the court confirmed the text as a legal person's work created by the law firm and affirmed the law firm as the proper

[9]The law firm sues the defendant for violation of the right of authorship, right of integrity and right of information communication on networks. See: *Beijing Film Law Firm v. Baidu* (2018) Beijing Internet Court Judgement Num. Jing0491 Min Chu No. 239 (京0491民初239号).

[10]Ibid.

[11]Ibid.

subject to file the lawsuit. There is also an ongoing debate among Chinese law professionals on whether CGWs can get copyright protection and if so what would be the protection model. Xiong (2017) supports to make use of the traditional copyright protection system, while others argue that CGWs can be regarded as a kind of derivative works,[12] which means there should be neighbouring rights instead of copyright over the outputs of computers.[13]

In conclusion, academia and courts hold different views on the legal status and protection models of CGWs. The core of the debate can be summarised into two main questions. One is about the copyrightability, or in detail, the identification of "work" in copyright law and the doctrine of originality. The other question is the relation between human authorship and CGWs. Both will be discussed in the following parts of this chapter.

3 The Copyrightability of CGWs Under Chinese Copyright Law

This section will discuss the copyrightability of CGWs under Chinese copyright law. To solve this problem, the author will first take a look at the concept of "work" in Chinese copyright law, followed by the analysis of originality, which is the concept applied to help determine "work" in copyright law.

When defining the word "work", Chinese copyright law does not emphasize human beings' participation in the process of creating a work. According to Article 3, of the Copyright Law of the People's Republic of China, only works belonging to the scope of literature, art and science, and works on the exhaustive list can be protected by the copyright law. The article states:

> ......the term "works" includes works of literature, art, natural science, social science, engineering technology and the like which are created in the following forms: (1) written works; (2) oral works; (3) musical, dramatic, qu yi and choreographic works; (4) Works of fine art and photographic works; (5) cinematographic, television and video-graphic works; (6) drawings of engineering designs and product designs, and descriptions thereof; (7) maps, sketches and other graphic works; (8) computer software; (9) other works as provided for in law and administrative rules and regulations.[14]

In the Regulations for the implementations of the copyright law of the people's republic of China, the definition of the "work" is:

---

[12]Yi (2017).

[13]Wei (2019).

[14]Article 3, the Copyright Law of the People's Republic of China. "qu yi" here means Chinese opera.

> the term "works" used in the law shall mean original intellectual creations in the literary, artistic and scientific domain, insofar as they are capable of being reproduced in a certain tangible form.[15]

The first paragraph of Article 5 of the ongoing draft version of Chinese copyright law[16] gives a new definition of the work:

> The term "work" refers to the original intellectual expression in the literary, artistic and scientific domain, insofar as they are capable of being fixed in a certain form.[17]

The legislation outlined above shows that works should belong to a specific subject matter, come from original intellectual creations (originality), and are able to be copied in a tangible form. Compared with the definition of "work" in the enacted regulations, the current proposal introduces two changes: from "intellectual creations" to "intellectual expression"; from "reproduced" to "fixed". Since the latter has less impact on the issue of copyright protection of CGWs, this article will primarily discuss the former.

Considering the definition of "work" in Chinese copyright law, the first question that needs to be explored is whether CGWs belong to a subject matter protected by copyright law. This means that the form of expression should conform to the scope and type of work that copyright law protects. In practice, many CGWs have the same form of expression as human authors' works. For instance, the poems of Xiaoice have been published as a book entitled *The Sunlight That Lost The Glass Window*. "The next Rembrandt" team made use of 3D printing technology to bring the new piece of computer-generated work to the canvas. Both works fall within the literary, artistic and scientific domain. These works can also be copied in tangible form. Accordingly, originality becomes the most controversial issue for the identification of CGWs.

With regard to the originality of CGWs, there are two main views in China. The first view is that computer-generated content can still be identified as "works", according to the current originality standards.[18] This supporting view emphasizes that existing CGWs are difficult to distinguish from human author's works.[19] It also recognizes that CGWs, based on artificial intelligence technology, are now different from previous computer-generated content.[20] It can deviate from the algorithm presets to solve new problems and generate new content.[21] However, this view does not go into great detail when looking at the originality of CGWs. For example, it fails to really answer questions such as: which part of the generation of

[15]Article 2, the Regulations for the implementations of the copyright law of the People's Republic of China (2013).

[16]The Exposure Draft of the Copyright Law of the People's Republic of China (2012).

[17]Article 5, the Exposure Draft of the Copyright Law of the People's Republic of China (2012).

[18]Xiong (2017).

[19]Xiong (2017).

[20]Xiong (2017).

[21]Xiong (2017).

the work can be considered as original? What about the creativity of computer-generated content? This viewpoint avoids directly responding to the criteria of the originality of CGWs. Other authors have put forward views regarding the originality of CGWs. They argue that the nature of artificial intelligence technology applied in the creative fields is computing rather than creation.[22] Considering the algorithm applied in the generation of content, the outputs of computers with the same materials could be highly similar.[23] The whole process of content generation is the lack of "intelligence" that is required in the creation of the work.[24] Thus, the originality of CGWs is difficult to establish. This view connects originality to "human beings and intelligence". Nonetheless, no human author or the actual intellectual contribution of human beings in the process of generation are not completely persuasive reasons to negate the originality of CGWs. The strength in this argument lies with the fact that artificial intelligence technology is not completely detached from human beings. It is coded by human beings and the process of application also contains some "human elements". Therefore, those two existing views seem to merit more exploration.

From the discussion above, the degree of creative participation of a person is the focus when identifying the originality of CGWs. Regarding the doctrine of originality, copyright law does not provide a specific definition of originality. However, according to legal practices and theories, the judgment standards for originality include the following. First, outputs of intellectual activities are the object of copyright law protection, as long as the work produced is the result of a certain degree of creativity. Looking back to the current legislation, according to Article 2, the Regulations for the implementations of the copyright law of the people's republic of China, "work" means original intellectual creations in a specific domain.[25] As for the concept of "creation", Article 3 of the Regulations provides that, "creation shall mean intellectual activities that directly produce literary, artistic, or scientific works."[26] Second, works should be created independently, and must not be copied or plagiarised.[27] Third, the "work" should achieve "the minimal threshold of originality".[28]As for the identification of the originality of CGWs, the point is still the clarification of the degree of human beings' participation. The level of human participation seems to be a key issue as to whether a work is original or not. Thus, the extent of human participation in the CGWs creation process needs further discussion.

[22]Wang (2017).

[23]Wang (2017).

[24]Wang (2017).

[25]Article 2, the Regulations for the implementations of the copyright law of the People's Republic of China.

[26]Article 3, the Regulations for the implementations of the copyright law of the People's Republic of China.

[27]Wang (2015).

[28]Wang (2015).

The process of content generation of CGWs is different from the creation of human author's works. As the output of algorithms, CGWs lack a certain sense of self-expression and communication, which is fundamentally different from the human creation. With artificial intelligence technology, a computer can discover the rules based on the analysis of big data and may choose a better solution in producing the result. This "ability" enables artificial intelligence to selectively capture relevant material and represent it without human participation, resulting in unpredictable content. Although this "learning" ability of computers can mimic human behavior to some extent, it may have limits due to the execution of the program, which is different from other essential abilities of human beings, such as aesthetics, intuition, imagination, etc.[29] Wei Lili, among others, have argued that the deep neural network algorithm of artificial intelligence is used to analyse the rules from the existing data information, and to refine and construct the model according to the user and that means the operation mechanism of deep learning ability is still related to human involvement.[30] However, the source of materials with a "human element" and the generation process with a human contribution are two different aspects. Moreover, it is difficult for CGWs to deviate from the path set by algorithms. Hence, there are basic differences between computer generation and human creation. With disputes over whether CGWs contain human personality, emotions, and ideas, the identification of CGWs as "works" becomes more complicated. In fact, the fundamental issue behind the discussion of the level of human participation in CGWs is how to factor in the role of the author's personality and creativity in determining originality. This issue requires further study.

To conclude, a creative production can be seen as a "work" under Chinese copyright law only when it belongs to a specific subject matter, has originality, and can be copied in tangible forms. For CGWs, many of which fall into the category of literary, artistic or scientific domains and can be copied in tangible forms can easily meet two of the three criteria. Therefore, the judgment of the originality of CGWs has become the key point of the copyrightability issue. Due to the difference between the process of computer generation and human creation, whether the author's personality and creativity have a substantial impact on the judgement of the originality of CGWs still requires further discussion.

4 The Authorship and CGWs in Chinese Copyright Law

Since the discussion of judgement criteria of originality is inseparable from the level of human contribution in CGWs, the next question is "Who is the author of CGWs?" The identification of authorship depends on the level of the creator's contribution. Looking back to the definition of "author", Chinese copyright law has clarified that

[29]Zhong (2017).

[30]Wei (2019).

"citizens who create works are authors".[31] The Article 3 of Regulations for the Implementation of the Copyright Law of the People's Republic of China provides the definition of "creation", and clarifies that "Any organizational activity, consultation, material support or other auxiliary services conducted or offered for another person's creation shall not be deemed as creation."[32] Nevertheless, how to measure human contribution in the process of computer generation is a difficult question. The analysis above has shown that there is a difference between the process of computer generation and human creation. Even though computers play a major role in content generation in CGWs, we cannot simply regard a computer as the author. In fact, the historic conception recognises that the author is often considered as "a creative, lone artist, inspired directly from nature",[33] which contains an assumption that the author is a human being. If computers account for the main contribution in content generation in CGWs, and they are not human beings, how to identify the author of CGWs?

Before discussing how to identify authors, a pre-question is that why do they need to be identified. This is because the very important function of identifying authorship is for confirming the ownership of copyright. The authorship is an indicator of the ownership of copyright. Article 11 (1) of the Chinese Copyright Law stipulates that "Except where otherwise provided in this Law, the copyright in a work shall belong to its author."[34] "Copyright belongs to the author" is the general principle of the ownership of Chinese copyright law. However, there are exceptions. With the development of the innovative industry and the commercialization of copyright trading, the role of investors is becoming more and more important. In order to protect the investment and enact certain policies in some specific areas, Chinese copyright law has made special provisions for ownership. Based on the general principle of "copyright belongs to the author", there are also some special rules such as "work created in the course of employment" and "the work of legal person or entity".[35] In general, the copyright of "work created in the course of employment" belongs to the author of the work. Article 16 of Chinese copyright law stipulates that "The copyright in such a work shall, subject to the provisions of the second paragraph of this Article, be enjoyed by the author."[36] However, in some circumstances, the legal person or other organization shall enjoy the copyright, excepting the right to claim authorship. Article 16 of Chinese copyright law indicates that specific types of works, which are created in the course of employment, mainly with the material and technical resources of the legal person or entity without legal personality and under its responsibility, or in accordance with law, administrative

[31] Article 11 (2), the Copyright Law of the People's Republic of China.

[32] Article 3, the Regulations for the implementations of the copyright law of the People's Republic of China.

[33] Maggiore (2018).

[34] Article 11 (1), the Copyright Law of the People's Republic of China.

[35] Article 16, Article17, the Copyright Law of the People's Republic of China.

[36] Article 16, the Copyright Law of the People's Republic of China.

rules and regulations or contracts concerned, ownership of copyright belongs to the legal person or entity without legal personality.[37] As for "the work of legal person or entity", the copyright law provides that "Where a work is created according to the will and under the sponsorship and the responsibility of a legal or entity without legal personality, such legal person or entity without legal personality shall be deemed to be the author of the work."[38] Both rules show the separation of the actual authors and the owner of the copyright and the separation of the act of creation and the copyright.

Under these specific situation, non-human entities such as corporate enterprises can be regarded as the author in some circumstances. Authorship therefore, can be separated from human beings, and existing regulations have gradually begun to attach more importance to protecting the interests of investors. Under the premise of confirming the copyrightability of CGWs, these rules which apply in special cases are possible approaches for the allocation of CGWs related rights. It also shows that there is space for Chinese copyright law to absorb CGWs under its current legal system. Even though the allocation of copyright of CGWs is not the primary question that requires to be solve at present, the current legal system can still provide possible approaches to fit into new situations, like those posed by AI-generated works.

In conclusion, authorship plays a vital role in copyright allocation. Generally, the author is the primary owner of the copyright of works. However, under specific situations, the authorship and the ownership are separated, which provides some possibilities for the current copyright legal system to interpret CGWs under its existing system and provide copyright protection.

5 Conclusion

AI technology is being widely applied to commercial uses in the innovative industries in China. This has led to academic and legal discussions concerning the availability of copyright for AI-generated works. Since AI-generated works are CGWs that are created without the participation of humans, their generation process

[37]Article 16, the Copyright Law of the People's Republic of China: The author of a work created in the course of employment in one of the following circumstances shall enjoy the right of authorship, while the legal person or entity without legal personality shall enjoy the other rights included in the copyright and may reward the author:

(1) drawings of engineering designs and product designs and descriptions thereof; computer software; maps and other works created in the course of employment mainly with the material and technical resources of the legal person or entity without legal personality and under its responsibility;
(2) works created in the course of employment where the copyright is, in accordance with laws, administrative rules and regulations or contracts concerned, enjoyed by the legal person or entity without legal personality.

[38]Article 11, the Copyright Law of the People's Republic of China.

differs from the traditional creative process, or the creative process for CAWs, where machines only play a role of tools to help human creators. This means that AI-generated works present a challenge for copyright law. However, Chinese copyright law, although lacking in precise regulations to cover these new kind of works, may contain a possible solution. According to the analysis above, whether the criterion of originality should take into account the author's personality and creativity in the context of CGWs is still controversial, and needs further discussion. However, some specific rules of the Chinese copyright law, such as "work created in the course of employment" and "the work of legal person or entity", provide possible approaches to absorb CGWs in the current legal system.

References

Cao Y (2016) The rationality of copyright protection for creation generated by artificial intelligence. Tech Law 3:488–508.曹源,《人工智能创作物获得版权保护的合理性》,《科技与法律》2016年第3期, 488–508页

Maggiore M (2018) Artificial intelligence, computer generated works and copyright. In: Bonadio E, Lucchi N (eds) Non-conventional copyright: do new and atypical works deserve protection? Edward Elgar, London

Tecent Research Institute (2017) The report of AI strategy in different countries. Technol Telecom 2:26–28. 腾讯研究院,《人工智能各国战略解读:联合国人工智能政策报告》,《电信网技术》2017年第2期, 26–28页

Wang Q (2015) Copyright Law. China Renmin University Press, Beijing, pp 20–27. 王迁:《著作权法》,中国人民大学出版社, 2015年, 第20–27页

Wang Q (2017) Qualitative research on the content of artificial intelligence in copyright law. Sci Law 5:148–155. 王迁,《论人工智能生成的内容在著作权法中的定性》,《法律科学》(西北政法大学学报)), 2017年第5期, 148–155页

Wei L (2019) Discussion on copyright issues of artificial intelligence products. J Zhengzhou Univ (Philos Soc Sci Ed) 3:22–26. 魏丽丽, 《人工智能生成物的著作权问题探讨》,《郑州大学学报》, 2019年第5期, 22–26页

Xiong Q (2017) Is Artificial intelligence generated content copyrightable? Intellect Prop 3:3–8. 熊琦,《人工智能生成作品的著作权认定》,《知识产权》2017年第3期, 3–6页

Yi J (2017) Is creation generated by Artificial intelligence work? Sci Law 5:137–147. 易继明,《人工智能创作物是作品吗?》,《法律科学(西北政法大学学报)》2017年第5期, 137–147页

Zhong Y (2017) Artificial intelligence: concept, approach and opportunity. Chin Sci Bull 22:2473–2479. 钟义信《人工智能:概念·方法·机遇》,《科学通报》, 2017年第 22 期, 2473–2479页

GPSR Compliance
The European Union's (EU) General Product Safety Regulation (GPSR) is a set of rules that requires consumer products to be safe and our obligations to ensure this.

If you have any concerns about our products, you can contact us on

ProductSafety@springernature.com

In case Publisher is established outside the EU, the EU authorized representative is:

Springer Nature Customer Service Center GmbH
Europaplatz 3
69115 Heidelberg, Germany

www.ingramcontent.com/pod-product-compliance
Ingram Content Group UK Ltd.
Pitfield, Milton Keynes, MK11 3LW, UK
UKHW021501020725
6662UKWH00056B/31
* 9 7 8 3 0 3 0 5 0 5 5 8 5 *